The Mar Chiquita Salt Lake (Córdoba, Argentina)

Enrique H. Bucher

The Mar Chiquita Salt Lake (Córdoba, Argentina)

Ecology and Conservation of the Largest Salt Lake in South America

 Springer

Enrique H. Bucher
IDEA Conicet and Centro de Zoología Aplicada
Universidad Nacional de Córdoba
Córdoba, Argentina

ISBN 978-3-030-15814-9   ISBN 978-3-030-15812-5   (eBook)
https://doi.org/10.1007/978-3-030-15812-5

This Springer imprint is published by the registered company Springer Nature Switzerland AG
The registered company address is: Gewerbestrasse 11, 6330 Cham, Switzerland

This book is dedicated to the memory of Hermann Frank, Helmuth Kanter, and Hans Seckt, who produced outstanding and pioneering work on the geology, limnology, and geography of Mar Chiquita during the first half of the twentieth century, at a time when this great lake was just a little known place in the center of Argentina.

Preface

Although the Mar Chiquita Lake is among the largest saline lakes of the world (2000–6000 km^2 area), little is known about the wetland outside Argentina, to a large extent due to its geographical isolation and the scarcity of scientific literature published in English, a deficiency that has started to be compensated in the recent years. Mar Chiquita has unique and interesting characteristics. The wetland is located in an area with subhumid climate instead of the arid environment characteristic of most salt lakes, which in turn determine wide variations in water level and salinity that has marked effects on the biota and local human population.

With this vision in mind, this book presents a systematization and integration of the knowledge accumulated to date on Mar Chiquita, covering a wide range of fields, including physical, biological, and historical aspects, and also environmental conservation and sustainable issues. The book is addressed to the specialists of the various fields covered, but is designed and written with the aim of making it accessible to the general public. Special attention has been given to the interconnection of the information provided in the different chapters, with an emphasis on dynamic and functional aspects. This publication results from the author work at the Mar Chiquita Biological Station in Miramar, Mar Chiquita, a research and environmental education unit created by the Universidad Nacional de Córdoba.

Córdoba, Argentina   Enrique H. Bucher

Acknowledgments

The author is grateful to the following institutions and people: Academia Nacional de Ciencias de Córdoba, Argentina, for supporting the publication of a previous version of this book in Spanish; the Academy former president, Alberto Maiztegui, for encouraging efforts to promote research and environmental education in Mar Chiquita; and also the US Fish and Wildlife Service's International Affairs and Neotropical Migratory Bird Conservation Act Programs for their financial support in research and conservation activities in Mar Chiquita. The author is indebted to his wife Elizabeth, for her permanent and dedicated support during the writing of this book.

Contents

Chapter 1
Geographic Overview

1.1 Introduction

The vast Mar Chiquita wetland, of about 10,000 km², is the final collector of the largest endorheic basin in Argentina, which includes parts of the provinces of Córdoba, Santiago del Estero, Tucuman, and Salta (Figs. 1.1, 1.2, and 1.3). Unlike most saline lakes that are found in desert areas (Hammer 1986), Mat Chiquita is located in a semiarid region with an annual rainfall of about 850 mm (Chap. 3). The lake was originally surrounded by the dry Chaco woodland, which has been deforested almost entirely and converted to agricultural land (Chap. 11). Most of the area is protected as a Multiple-Use Reserve by the Province of Córdoba (Fig. 1.4, Chap. 13).

1.2 Subregions

The Dulce River wetland includes two hydrologically integrated subregions: the Rio Dulce wetland (locally known as *Bañados del Rio Dulce*) and the Mar Chiquita Lake (locally known as *Laguna Mar Chiquita o Mar de Ansenuza*) (Figs. 1.2 and 1.3). The Dulce River wetland is generated by the terminal floodplain of the Dulce River, on the northern coast of the Mar Chiquita Lake (Fig. 1.4). The dominant landscape is complex and heterogeneous, including the Dulce River meandering stream, temporary and permanent ponds, extended grasslands, halophytic shrublands, and elevated areas with woody vegetation (Chap. 10).

The Mar Chiquita Lake, with an area that at maximum level covers about 6000 km² and a length of 100 km east-west and 80 km north-south, is the largest salt lake in South America and among the largest in the world. The lake is shallow in most of its area, with the deepest sites reaching about 10 m at the highest lake level.

© Springer Nature Switzerland AG 2019
E. H. Bucher, *The Mar Chiquita Salt Lake (Córdoba, Argentina)*,
https://doi.org/10.1007/978-3-030-15812-5_1

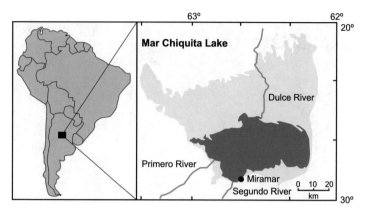

Fig. 1.1 Location of the Mar Chiquita wetland in South America. On the right, the dark blue color corresponds to the Mar Chiquita Lake area during the lowstand period (before 1980). The gray area indicates the lake's greatest expansion area during the highstand period after the 1980s

Fig. 1.2 Drainage basin of Mar Chiquita. (1) Sierras Pampeanas Range (Aconquija mountains), (2) Salado River, (3) Rio Hondo dam, (4) Dulce River, (5) Ambargasta salt flats, (6) Saladillo branch of the Dulce River, (7) Western dispersed creeks, (8) Rio Dulce wetland, (9) Sierras Pampeamas range (Córdoba mountains) de Córdoba mountains, (10) Primero River, (11) Mar Chiquita Lake, (12) Segundo River. Dotted lines indicate provincial limits

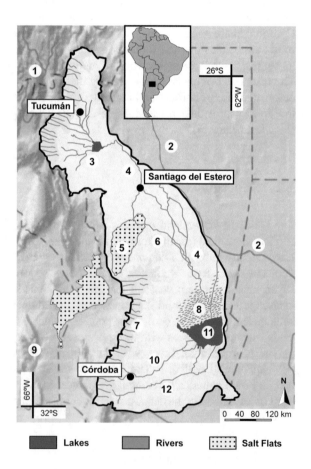

Fig. 1.3 Satellite MODIS image of Mar Chiquita and the northern Dulce River wetland showing an incipient stage of the annual flood of the wetland

Seen from the air, the lake looks like an inland sea, to the point that it attracts marine bird species that are usually not seen far from the oceanic coasts (Chap. 7).

1.3 An Always Fluctuating Lake

One of the most distinctive characteristics of this lake is its high variability in terms of size, water level, and salinity over time. Contrarily to most saline lakes of the world that exhibited a declining trend in level during the twentieth century (Wurtsbaugh et al. 2017), Mar Chiquita has shown a dramatic increase in water level since the late 1970s, due to a sustained increment in rainfall in the catchment area, which is probably related to global climate change (Chap. 3). Accordingly, since the middle of the twentieth century to the present, the lake has ranged between the extreme hypersaline condition, with salinity above 100 g/L, through hypersaline (50–100 g/L), to mesosaline (20–50 g/L) condition, and back to hypersaline condition (Chap. 4).

This expansion-retreat pulse brought several significant (and unexpected) changes in the whole wetland. Among them, the most significant were the flooding of the only coastal town (Miramar) (Fig. 11.1, Chap. 11) and the invasion of the lake

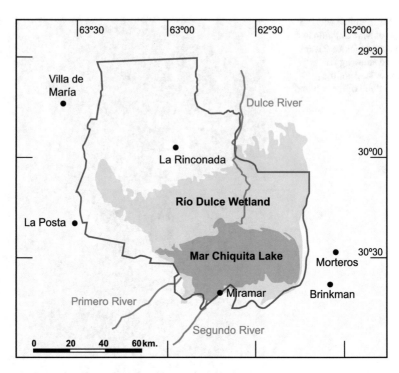

Fig. 1.4 The limits of the Mar Chiquita protected (Reserva de Uso Multiple de la Provincia de Córdoba) (red line) and the main urban centers in the region (black dots). The blue color indicates the Mar Chiquita Lake size in the lowstand period, and the gray area shows the maximum lake expansion attained during the highstands

by the pejerrey silverside fish (*Odontesthes bonariensis*) (Chaps. 4, 5, and 11). Both events had never been recorded before. In addition, vast saline playas were left exposed by the expansion and retreat of the water edge, where huge salt-dust storms were generated (Chaps. 2 and 3).

1.4 Biodiversity Richness and Conservation

The wetland provides a suitable habitat to impressive biodiversity (Chaps. 5, 6, 7, 8, and 9). Birds are particularly common and diverse. For this reason, the area was declared a reserve by the Province of Córdoba and a Ramsar site by the Ramsar Convention for the protection of wetlands. The area harbors outstanding numbers of migrating and resident water birds, including three of the six extant flamingos in the world (Fig. 1.5) and several species of transcontinental migratory shorebirds. The wetland is also the habitat of several globally endangered species (Chaps. 7 and 13).

Fig. 1.5 Partial view of a large Chilean flamingo breeding colony in the Mar Chiquita Lake of about 15,000 breeding birds. (From Bucher 2006)

Main significant threats to the site include agriculture, deforestation, and unregulated tourism. There is a high risk that the unregulated extraction of water for irrigation and other purposes from the tributaries, and particularly from the Dulce River, could markedly reduce or even totally dry out the lake. Growing water pollution from local urban and industrial sources is becoming an important issue (Chap. 13).

In the Dulce River wetland, transhumant pastoralism was dominant until recently. At present, a shift to sedentary, highly technified beef production systems is developing rapidly, creating significant environmental and social conflicts (Bucher 2016), (Chap. 10).

1.5   Human Presence in the Region: A Historical Review

Several native groups inhabited the Mar Chiquita wetlands area before the arrival of the Spanish conquerors, living on hunting, fishing, and incipient agriculture. The wetland was discovered by the Diego de Rojas expedition coming from Lima, Peru, in 1545. The whole wetland remained under the control of the natives for about three centuries, given that the low agricultural potential of the soil and the frequent raids of hostile natives discouraged settlers from establishing in the region. The effective occupation of the wetland only started after 1860, when the Argentine army gained control of the entire Chaco region (Bucher et al. 2006), (Chap. 11). At

Fig. 1.6 Aerial view of the city of Miramar, the only urban settlement along the Mar Chiquita Lake coast. This rapidly growing resort was partly destroyed by a flood during a high wáter period at the end of the twentieth century (Chap. 3)

present, the Mar Chiquita wetland is surrounded by a well-populated area, with a dense network of roads and urban centers. In contrast, the human population in the wetland is low and very sparse with only one small town in the region (La Rinconada: −30.183S, −62,947 W).

With regard to the Mar Chiquita Lake, the city of Miramar (−30.919 S, −62.677 W), with about 4500 inhabitants, is the only urban center on the entire lakeshore (Fig. 1.6). Located on the south coast, the city is the main tourism center for the whole Mar Chiquita region. During the highstand period, between 1977 and 2003, the city of Miramar suffered two flooding episodes that affected about half of the urban area, including downtown (Fig. 1.7).

1.6   On the Name of Mar Chiquita

The name of this Mar Chiquita Lake has changed over time. The first name that appears in South American maps after the discovery of the continent in the seventeenth century was "*Lagunas saladas*" or "*Lagunas saladas de los porongos*"

Fig. 1.7 The city of Miramar partly flooded in 1984 during the highstand period

(meaning "salt lagoons" or "porongos salt lagoons," respectively) (Bucher et al. 2006) (Chap. 11).

The word *porongos* (singular: *porongo*) originates from the word *poronko* in the South American Quechuan language. This term refers to wide, hemispheric, ceramic jars frequently found in the archaeological sites in the region (Aparicio 1942). A much less likely alternative translation for the term "*porongo*" refers to the small fruit of the pumpkin plant bottle gourd (*Lagenaria siceraria*), widely used by the native peoples throughout South America for drinking a kind of green tea known as *yerba mate* (*Ilex paraguariensis*).

The name Mar Chiquita appears for the first time early in the eighteenth century (1811) in a map used by the Argentine army during the independence war against Spain, ordered by the General Martin de Pueyrredon (Bucher et al. 2006). Later, this name was used by De Moussy (1861) in a book on the general geography of Argentina. Since then, this name has been used as the official name of the lake until present. It is worth mentioning that there are two more lakes named Mar Chiquita ("Little Sea") in Argentina, located in the province of Buenos Aires, eastern Argentina. One is a lagoon adjacent to the town of Junín (34°26′ S 61°10′ W), and the other is a coastal lagoon on the Atlantic Ocean (37°37′ S 57°24′ W) near the city of Mar del Plata.

Later in the twentieth century, Cabrera (1931) published a review of the historical documents related to Mar Chiquita, in which the author concludes that the original name was "Ansenuza." Further research has shown that the name Ansenuza did not correspond to the lake itself, but only to a portion of land close to the southern coast of the lake (Montes 2008; Bucher et al. 2006). At present, both names were integrated into a new designation: "Laguna Mar Chiquita de Ansenuza."

References

Aparicio F (1942) Arqueología de la laguna de Los Porongos. Relaciones de la Sociedad Argentina de Antropologia 3:45–52

Bucher EH (2006) Bañados del Rio Dulce y Laguna Mar Chiquita. Academia Nacional de Ciencias (Córdoba, Argentina), Argentina

Bucher EH (2016) El futuro incierto de los humedales del Chaco: el caso de los bañados del Rio Dulce. Paraquarua Nat 4(11):18

Bucher EH, Marcellino A, Ferreyra C, Molli A (2006) Historia del Poblamiento Humano. In: Bucher E (ed) Bañados del Rio Dulce y Laguna Mar Chiquita (Córdoba, Argentina). Academia Nacional de Ciencias (Córdoba, Argentina), Córdoba, pp 301–325

Cabrera JP (1931) Ensayos sobre etnología argentina. Peuser, Buenos Aires

De Moussy VM (1861) Description Geographique et statistique de la Conferderation Argentine. Firmin-Didot, Paris

Hammer UT (1986) Saline Lake ecosystems of the world. Dr. W. Junk Publishers, Boston

Montes A (2008) Indígenas y conquistadores de Córdoba. Ediciones Isquitipe, Buenos Aires

Wurtsbaugh WA, Miller C, Null SE, DeRose RJ, Wilcock P, Hahnenberger M, Howe F, Moore J (2017) Decline of the world's saline lakes. Nat Geosci 10:816–821

Chapter 2
Geomorphology

2.1 Introduction

The Mar Chiquita Lake is a large inland lake in central Argentina (between 26–32°S and 62–66°W). The lake is the depocenter of the largest quaternary endorheic basin in Argentina, which includes parts of the provinces of Cordoba, Santiago del Estero, Tucuman, and Salta. The main tributaries are the Primero, Segundo, and Dulce rivers (Fig. 2.1).

This chapter includes an overview of the drainage basin main characteristics and a detailed description of the Mar Chiquita Lake geomorphology and its geological history. Two additional aspects relevant to Mar Chiquita include (a) the displacement of the tributary rivers course in historical times and (b) the generation of enormous salt dust storms over Mar Chiquita derived from recent climate changes in the region.

2.2 The Drainage Basin

This large basin is part of the Chaco-Pampas Plain, a geological region that extends from the central and northern lowlands of Argentina to the Paraguayan and Bolivian Chaco. At a larger scale, the Chaco-Pampas Plain is the southern portion of the vast South American deposition trough that extends from the Venezuelan plains in the north to the Río de la Plata region in the south (Kruck et al. 2011) (Fig. 2.1). Most of The Mar Chiquita basin was originally occupied by a woodland savanna known as the Chaco forest (Bucher 2006). At present, a large portion of the area has been deforested and dedicated to agriculture and cattle raising (Cabido and Zak 1999; Kruck et al. 2011). The hydrological characteristics of the basin are detailed in Chap. 5.

© Springer Nature Switzerland AG 2019
E. H. Bucher, *The Mar Chiquita Salt Lake (Córdoba, Argentina)*,
https://doi.org/10.1007/978-3-030-15812-5_2

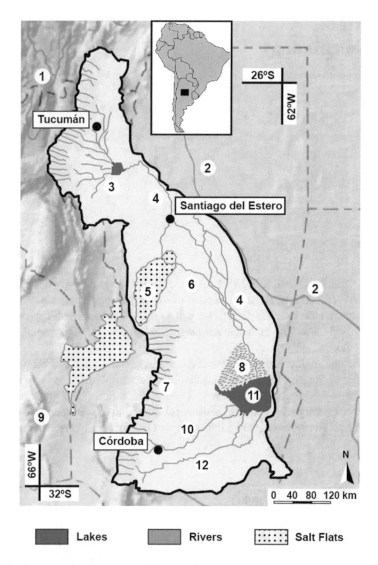

Fig. 2.1 (1) Aconquija mountains, (2) Salado River, (3) Rio Hondo dam, (4) Dulce River, (5) Ambargasta salt flats, (6) Saladillo branch of the Dulce River, (7) Western dispersed creeks, (8) Rio Dulce wetland, (9) Sierras Pampeanas Range (Córdoba mountains), (10) Primero River, (11) Mar Chiquita Lake, (12) Segundo River. Dotted lines indicate provincial limits

2.3   The Mar Chiquita Lake

The Mar Chiquita Lake is located in the intersection of two large geomorphological regions: the Sierras Pampeanas Range and the Pampas Plain (Fig. 2.1). The Sierras Pampeanas Range (locally known as Sierras de Córdoba), located to the west, is an old crystalline massif that has been subjected to successive erosive events, forming a landscape characterized by relicts of erosional surfaces, fractured and tilted during the Andean movements that started ~50 to 40 million years ago (Carignano 1999).

This group of mountains presents an abrupt western flank that corresponds to large fault scarps and an extended oriental slope where the old erosion surfaces are preserved, sometimes covered by tertiary and quaternary sediments (Piovano et al. 2009) deposited by a series of coalescent alluvial fans. These fluvial sediments become finer toward the east, where they gradually mix with the Aeolian sediments of the loessic Pampas Plain (Cioccale 1999). The Aeolian sediments are classified as typical loess with grain size dominated by coarse silt. They are composed of metamorphic and igneous rocks and volcanoclastic material belonging to the Andes Range, Sierras Pampeanas Range, and Paraná basin Aeolian materials that were accumulated by southern winds in the Pampean Aeolian System (Brunetto and Iriondo 2007).

The lake occupies a sunken landform that is contained from the west, by the basement-cored Sierras de Córdoba, and to the east, by a buried westward-verging thrust fault, the Tostado-Selva fault, which forms a broad topographic swell known as the San Guillermo High (Fig. 2.2). To the south, it ends on the Pampas Plain. To the north, it opens onto the Chaco woodland plains of the Santiago del Estero province (Kruck et al. 2011).

2.3.1   The Lake Geomorphology

Lake Coasts   The Mar Chiquita Lake has an extensive littoral belt. The entire lake coasts are sedimentary shores, including the vast exposed shores of very low slope on the northern coast, which rise gradually into the wide Rio Dulce delta. They become wide salt flats in dry years. Moderately exposed sandy shores are common along the rest of the lake, except on the southwest of the lake, where exposed sandy dune shores are prominent, associated with eroded sand fields and active and dissipated fossil dunes. In some places they form exposed sandy cliff shores with scarps of considerable height (10–20 m) and length of about 500 m (Kröhling and Iriondo 1999) (Fig. 2.3). Most of the lake coasts were originally covered with the native Chaco woodland, which at present has been replaced by agricultural land.

River Deltas   The Dulce River delta on the northern coast and the Segundo River (Plujunta) on the southern coast are the most prominent. The Primero River flows in the Laguna del Plata lagoon. Of particular interest is the Segundo Viejo River, an old riverbed branch of the Segundo River that opens on the southeast corner of the lake in a partly eroded, complex delta that provides habitat to a wide diversity of wildlife, particularly birds (Fig. 2.4).

Fig. 2.2 Upper image:
Mar Chiquita wetland: (1)
Chaco woodland
ecoregion, (2) Rio Dulce
wetland, (3) San Guillermo
High, (4) coastal sand
dunes area. Red line
indicates the Tostado-Selva
fault. Lower image.
Southern Mar Chiquita
Lake coast. (5) Laguna del
Plata lagoon, (6) Segundo
(Plujunta) River delta, (7)
Miramar City on the coast,
(8) Primero River, (9)
Segundo (Plujunta) River

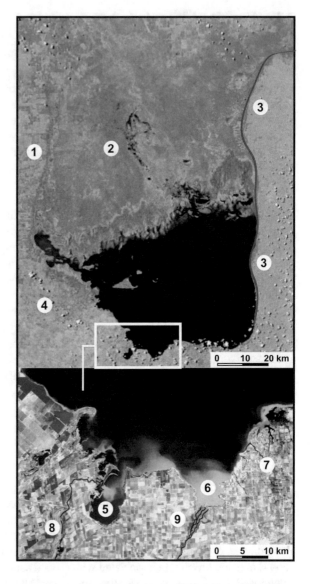

Laguna del Plata On the southwestern coast of the lake, the Primero River flows into an 18 km^2 lagoon (known as *Laguna del Plata*, and also *Lagunilla del Plata*), which in turn connects to the main lake through a short, relatively narrow channel during the low stands, becoming a lake's bay during high water level periods (Fig. 2.2, lower image).

Islands Many islands emerge during the low-stand period, most of which disappear during the lake highstands. The only exception is the large El Mistolar island on the western coast, a portion of a degraded mega-dune that becomes isolated from

Fig. 2.3 Coastal fossil sand dunes on the southwest coast of the Mar Chiquita Lake

the continent during the highstand periods. Another distinctive island is a 30-km-long barrier island that runs parallel to the eastern half of the north coast, remaining emerged while the water level is below 68 m a.s.l. This barrier is the remaining of a high coast with a dense Chaco forest that was subject to intense timber exploitation before being flooded during the highstand period (Bucher et al. 2006b).

2.3.2   The Lake Bathymetry and Sediments

The lake is very shallow, with a maximum depth of about 10 m at the highest recorded water level. The lake bottom has a slight north-south slope, with the deepest area located close to the southern coast (see Chap. 5 for details on the lake bathymetry).

Lake Bottom Sediments The sediments at the bottom of Mar Chiquita Lake are heterogeneous due to the lake size, the large amounts of material deposited by the incoming rivers, and the expansions and retreats of the lake coast over time caused by the long-term variations in water level (see Chap. 3). The average proportion of the dominant minerals (weight) is 39% clay, 47% silt, and 14% sand (Martinez 1995). Clay minerals consist of abundant illite, with lesser amounts of calcium-rich

Fig. 2.4 The terminal portion of a no longer active branch of the Segundo River known as Rio Segundo Viejo. Located on the southeast of the lake, close to the locality of Altos de Chipion, it generates several semi-closed lagoons with a gradient in salinity and vegetation that provides suitable habitats for a very diverse bird fauna

smectite and minor amounts of kaolinite. Calcite is the dominant carbonate mineral (Martínez 1995; McGlue et al. 2015) (Chap. 4).

From the point of view of their origin, the lake sediments can be grouped into three basic categories: (a) materials found on the vast areas where the coastline expanded substantially after the highstand period starting at the end of the 1970s (pedogenic), (b) inputs from the tributary rivers (allogenic), and sediments generated in the lake (authigenic) (Martinez 1995).

Pedogenic Sediments These are found particularly on the northern lake coast, where the lake coastline expanded substantially during the highstand period. These sediments are characterized by relatively high silica and aluminum content, low levels of calcium oxide and carbonates, and lack of gypsum and calcite crystals (Martinez 1995). Pedogenic sediments are also found in submerged dune fields on the southwest of the lake (Kröhling and Iriondo 1999).

Allogenic Sediments These originate in the deltas generated at the mouth of the Dulce, Segundo, and Primero rivers and at the paleodeltas of the Segundo Viejo River. Most of the sediments brought by the Primero River remain trapped in the

Laguna del Plata lagoon before reaching the open lake (Fig. 2.2). Most of the coarse material transported by the rivers consists of sand and silt. The mineralogy of this material includes quartz, plagioclase, volcanic glass, and micas (McGlue et al. 2015). Most of the remaining allogenic sediments are composed of fine fragments of clay and mud. Organic matter is another important component of the allogenic material, including living organisms such as fish, most of which cannot survive the sudden change from freshwater to saline water conditions. Oxidation and reworking of organic matter are common in these types of high-energy environments.

Authigenic Sediments   These are found in the south-central depression of the lake, in the deepest sectors. It is very likely that these sediments have been submerged since the origin of Mar Chiquita. They are characterized by the dominance of clays, low silica and aluminum content, and relatively high calcite and gypsum values (in large crystals). They also contain sulfides, ammonium, and organic mud, indicating anoxic conditions (McGlue et al. 2015; Martinez 1991).

The authigenic organic-rich mud is an important natural feature of Mar Chiquita Lake, both because of its peculiar characteristics and origin (Martínez 1991; McGlue et al. 2015) and also due to its value as a health product, which promotes tourism in the region. The formation of organic mud is a common feature in saline lakes, and in most of them, it is considered of therapeutic value (Hammer 1986). In terms of its mineral components, these profundal muds contain ~33% clay minerals, 10% calcite, 7% quartz, and 4.5% halite plus gypsum, with the remaining portion composed of feldspars, altered volcanic glass, and amorphous material (organic matter and diatoms) (McGlue et al. 2015) (Chap. 4).

2.3.3   The Lake Geological History

The Chaco-Pampas Plain was traditionally considered a stable area, but recent studies showed that it was affected by neotectonic activity and that Andean deformations reached regions located more than 700 km from the Andean chain. According to Mon and Gutierrez (2009), one of these deformations may have originated the activation of the Tostado-Selva fault and the uplift of the San Guillermo High (Figs. 2.1 and 2.2).

As a result, the Dulce River was impounded to flow southward along the San Guillermo High fault until joining the Tercero River, the only river originating in the Sierras de Córdoba Mountains that reaches the Parana River. This old connection is still evident when the old riverbed becomes flooded under high rainfall episodes (Fig. 2.5).

In addition, Mon and Gutierrez (2009) consider that the Guillermo High modified the flow of the Primero and Segundo rivers, which turned northeast their previous eastward-flowing. In time, the accumulation of the alluvial fans generated by the two rivers on the western flank of the San Guillermo closed the southern flow of

Fig. 2.5 MODIS satellite image (March 3, 2014) during a high rainfall period. (1) Lower Dulce River; (2) area of close contact between the Salado River and an old branch of the Dulce River, which allowed episodes of avulsion of the Salado River flowing into the Dulce River; (3) sub-meridional lowlands wetland in Santa Fe province; (4) San Guillermo High; (5) Mar Chiquita Lake; (6) the flooded old Rio Dulce riverbed south of Mar Chiquita, which connects with the Tercero River; (7) Paraná River; (8) Tercero River. Image from MODIS Aeronet Cordoba CETT_2015 03 14 terra_367 bu_250m

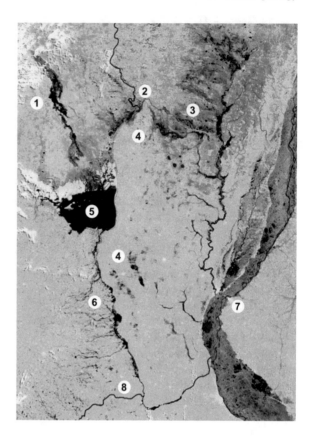

the Dulce River, generating the Mar Chiquita Lake. The authors estimate that the hydrologic lake formation took place during the Middle Pleistocene, although they agree on the lack of absolute chronology data to support their hypotheses. They also mention that the existence of the Dulce River valley southward of Mar Chiquita clearly indicates that lacustrine conditions are younger than the uplift of the San Guillermo High, which is dated at the Middle Pleistocene as mentioned before.

2.3.4 Mar Chiquita Paleoclimates

Kröhling and Iriondo (1999) mapped the Mar Chiquita geomorphology in detail, and, according to sedimentological properties of the quaternary deposits, they estimated climatic events in the area. Cioccale (1999) used several lines of evidence to explain the climatic fluctuations that occurred in the central region of Argentina during the last millennium. Another important contribution to the reconstruction of the paleoenvironmental evolution of the Late Pleistocene and Holocene in Mar Chiquita region has been produced by Piovano et al. (2009) based on sedimentary cores retrieved from the Mar Chiquita Lake, interpretation of satellite images, meteorological records, historical documentary sources, and fieldwork.

In addition, complementary information along this line of research can be found in Coianiz et al. (2014) and Piovano et al. (2004a, b, 2009). The existing information suggests that past climatic conditions were predominantly dry, with relatively short humid periods in between, with no indication that rainfall could have reached higher values than those recorded during the recent highstand period (Piovano et al. 2006).

2.4   Tributary Rivers and Their Recent Course Changes

At present, three rivers bring water to the Mar Chiquita wetland. The Dulce River comes from the north and the Primero and Segundo rivers from the south (Fig. 2.1). Both the Rio Dulce and the Salado rivers flow across the Gran Chaco region in parallel courses from NW to SW. While the Dulce River ends in Mar Chiquita, the Salado River flows into the Parana River (Fig. 2.1). In the past, the Salado River connected intermittently to the Dulce river (and therefore to Mar Chiquita) at least since the seventeenth century. At present both flow in parallel courses along the Chaco plain covered by fluvial and Aeolian Late Pleistocene and Holocene sediments. The Primero and Segundo rivers lower reach flow through the very flat Pampas Plain (Kröhling and Iriondo 1999; Kruck et al. 2011).

All these rivers have gone through important river course changes in recent centuries. River avulsion has been a common process in the very shallow plains of the final portions of the Mar Chiquita tributary rivers and, in some cases, reached great magnitude (Bucher et al. 2006a). The well-known tendency of the Chaco and Pampas rivers to frequent course change led one of the early European travelers in the region to call them "itinerant rivers" (Carrió de la Vandera 1908). A detailed description of the Mar Chiquita tributaries and their historical changes follows, based mostly on Bucher et al. (2006a).

Dulce River This river had a main change of course in 1825, when a major flood led to the partial avulsion of the river near the village of Sumamao, in the Santiago del Estero province. As a result, most of the flow diverted to the south, leaving only a small flow on the old bed. The new course crossed the vast salt plain of the Salinas de Ambargasta, where water takes up salts. After leaving the salt flat, it takes the name of Saladillo and finally joins the ancient course of the Dulce River again (Fig. 2.1). About 100 years later (1933), a new great flood caused the Rio Dulce to retake its original channel and abandon the Saladillo, which since then is reduced to a small channel that only carries water during years of high precipitation (Soldano 1947).

Salado River It is one of the major continental-scale rivers in South America. The Salado River had two total avulsion episodes by which it deviated to the west and started flowing into the low Dulce River at the Laguna de Los Porongos site, at a short distance from the Mar Chiquita Lake. The first occurred between 1647 and 1655 and the second from about 1758 until the end of the seventeenth century (Jolis 1972; Soldano 1947; Roverano 1955; Dussel and Herrera 1999). During these periods, the Mar Chiquita Lake must have received a significant addition of water inflow, which probably had considerable hydrological and biological effects.

At present, the occurrence of sporadic and limited water exchanges between the Dulce and Salado Rivers continue to be a common event, during exceptional overflows of any of both rivers. The water flow usually takes place in an area where both rivers run very close (−29.407514S, −61.927054W), near the city of Tostado in the province of Santa Fe.

The close connection between the two rivers may have had (and still has) a significant influence, not only in hydrological terms but also from the biogeographic perspective, providing a corridor for fish and other species between the Paraná-Paraguay basin and the Dulce River basins (Chap. 6). The connecting wetlands also allow terrestrial species movements between the Rio Dulce wetland and the large wetland known as *Bajos Submeridionales* (sub-meridional lowlands) eastward from Mar Chiquita in the Santa Fe province (Lewis 1993) (Fig. 2.5). There is abundant evidence showing that several species may move into Mar Chiquita during the highstand periods and became temporarily extinct during the lowstands (see Chap. 8).

Segundo River The whole of the lower reach of the Segundo River was characterized by a very flat land with a succession of wetlands and lagoons up to the river mouth on the lake, where a wide delta was generated. Of particular importance was a large forest swamp known as "Bañados de El Tio," at about 70 km before reaching the southeastern corner of the lake in a wide valley with a complex wetland landscape known as Saladillo or Segundo Viejo (Kanter 1932). With the purpose of drying the El Tio swamp, in 1927 the Segundo River was deviated and canalized into the Plujunta creek, which opened into the lake close to Miramar town (Ninci 1919) (Fig. 2.2). Since then, Río Segundo Viejo no longer contains active channels. However, it remains as an important wetland with a sunken delta (Fig. 2.4).

Primero River In its final stretch, the river presents two arms called new and old (Fig. 2.2). The latter had greater water flow until 1886 when an exceptional flood diverted the course toward the new arm, which continues active up to the present. In addition, information from prehistoric times indicates that the Primero River had other courses opening more to the west, in a west to east displacement, following the general land slope in the region (Frenguelli and De Aparicio 1932) (Fig. 2.2).

2.5 Salt Dust Storms over Mar Chiquita

The exceptional changes in the Mar Chiquita wetland induced by the dramatic increase in water level that occurred since the 1970s led to the expansion of the lake area to a maximum of about 7000 km² in 2003 and a subsequent reduction during the

Fig. 2.6 Large portions of the Mar Chiquita northern coast were flooded during the highstand period. The original Chaco woodland was completely eliminated. When the water level receded, the area became a large salt flat from where salt dust storms develop. Isolated dead trees are the last testimony of the original landscape

following 2003–2014 dry period that left about 4500 km² of exposed dry bottom mudflats where all the original vegetation was eliminated (Fig. 2.6). As the lake shrunk, the area occupied by salt playas expanded, and the number of salt dust storm events grew accordingly (Figs. 2.7 and 2.8). The intensity and frequency of salt dust storms reached an exceptional size in some years, extending over 800 km (Fig. 2.7).

The frequency of events correlated with the size of the salt mudflats (Fig. 2.8). Interestingly, the salt dust storms were almost entirely restricted during the cold months of the year (May to September), with a clear peak in the coldest months (June, July, and August). The mean monthly temperature during the latter period was below 10 °C (Fig. 2.9). The observed restriction of salt dust storms to cold weather is related to the fact that the underground water below mudflats is rich in sodium sulfate. When this solution emerges due to capillarity, the salt may crystallize in the form of hydrous crystals known as *mirabilite*, but only when under low temperature. Once these *mirabilite* crystals are deposited on the saline playas in the form of "fluffy," highly erodible crystal, they can easily be lifted and dispersed by winds of sufficient intensity (Bucher and Stein 2016).

The atmospheric transport and dispersion of the salt dust transported by wind were estimated using HYSPLIT, a widely used modeling system, which integrates the effect of the emissions, transport, dispersion, and deposition based on the climatic conditions in a given year (Bucher and Stein 2016).

The HYSPLIT simulation for the year 2009 estimated a salt dust deposition maximum near the sources of about 2.5 kg/ha/year and a decreasing trend from the emission area outward. During that year, the total deposition generated in Mar Chiquita was estimated at 6.5 million tons. Even if smaller than the 43 million tons

Fig. 2.7 MODIS image of a powerful salt dust event generated in Mar Chiquita. Date: September 10, 2013. Distance reached by the visible plume: about 800 km. Modis Image: AERONET Cordoba-CETT (aqua), 250 m pixel size, full. Color. doi: https://doi.org/10.1371/journal.pone.0156672. g006 PLOS

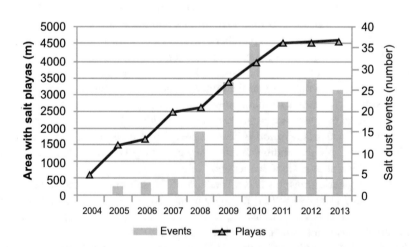

Fig. 2.8 Area occupied by salt playas and number of salt dust events recorded in Mar Chiquita during 2004–2013 period using MODIS images. (From Bucher and Stein. doi: https://doi.org/10.1371/journal.pone.0156672.g005)

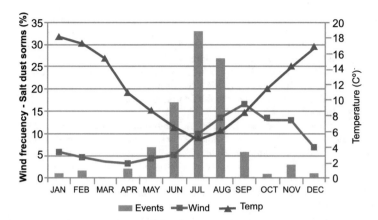

Fig. 2.9 Mar Chiquita: monthly frequency of salt dust storms (%), days with wind over 27 km/h (%), and mean minimal monthly temperature (C). Temperature records from Miramar, 30° 57'S; 56° 20' W. Servicio Meteorológico Nacional, Argentina. (From doi: https://doi.org/10.1371/journal.pone.0156672.g007 PLOS)

found for the Aral Sea, it is still significant and proportionally similar if we consider the smaller size of Mar Chiquita. Moreover, the massive dust plumes generated in Mar Chiquita (over 500 km long and 90 km wide) are of comparable magnitude with those recorded in the Aral Sea (Bucher and Stein 2016).

It is worth mentioning that the massive dust storms in Mar Chiquita have the distinctive characteristics of not being caused by lake shrinking due to water diversion in the upper tributaries, as is the usual case with other saline lakes like the Aral Sea. Instead, the storms resulted from a large-scale, 30-year pulse of expanding and receding waters due to drastically changing rainfall conditions, of a magnitude not recorded before since the lake origin in the Middle Pleistocene (Kröhling and Iriondo 1999; Piovano et al. 2006).

References

Brunetto E, Iriondo MH (2007) Neotectónica en la Pampa Norte (Argentina). Rev Soc Geol Esp 20:17–29

Bucher EH (ed) (2006) Bañados del Rio Dulce y Laguna Mar Chiquita (Córdoba, Argentina). Academia Nacional de Ciencias (Córdoba, Argentina), Córdoba, Argentina

Bucher EH, Stein AF (2016) Large salt dust storms follow a 30-year rainfall cycle in the Mar Chiquita Lake (Córdoba, Argentina). PLoS One 11(6):e0156672

Bucher EH, Marcellino A, Ferreyra C, Molli A (2006a) Historia del Poblamiento Humano. In: Bucher E (ed) Bañados del Rio Dulce y Laguna Mar Chiquita (Córdoba, Argentina). Academia Nacional de Ciencias (Córdoba, Argentina), Córdoba, pp 301–325

Bucher EH, Gavier Pizarro G, Curto ED (2006b) Sintesis geografica. In: Bucher EH (ed) Bañados del rio Dulce y laguna Mar Chiquita (Córdoba, Argentina). Academia Nacional de Ciencias (Córdoba, Argentina), Córdoba, pp 15–27

Cabido M, Zak M (1999) Vegetación del norte de Córdoba. Secretaria de Agricultura, Ganaderia y Recursos Renovables de la provincia de Córdoba, Argentina, Córdoba

Carignano CA (1999) Late Pleistocene to recent climate change in Córdoba province, Argentina: geomorphological evidence. Quat Int 57–58:117–134

Carrió de la Vandera AC (1908) El lazarillo de ciegos caminantes desde Buenos Aires hasta Lima (1773). Biblioteca de la Junta de Historia y Numismática Americana, Buenos Aires

Cioccale M (1999) Climatic fluctuation in the Central region of Argentina in the last 1000 years. Quaternary International, 62

Coianiz LJ, Ariztegui D, Piovano E, Lami A, Guilizzoni P, Gerli S Waldmann N (2014) Environmental change in subtropical South America for the last two millennia as shown by lacustrine pigments. J Paleolimnol, 53. https://doi.org/10.1007/s10933-014-9822-2

Dussel P, Herrera R (1999) Repercusiones socioeconó- micas del cambio de curso del Río Salado en la segunda mitad del siglo XVIII. In: Garcia Martinez B, Gonzalez JA (eds) Estudios sobre historia y am- biente en América. El Colegio de México. Instituto Panamericano de Geografía e Historia, Ciudad de México

Frenguelli J, De Aparicio F (1932) Excursión a la Laguna de Mar Chiquita (Provincia de Córdoba). Publicaciones del Museo Antropológico y Etnográfico de la Facultad de Filosofía y Letras. Universidad de Buenos Aires, Seria A 2, pp 121–132

Hammer UT (1986) Saline Lake ecosystems of the world. Dr. W. Junk Publishers, Boston

Jolis J (1972) Ensayo sobre la historia natural del Gran Chaco. Universidad Nacional del Nordeste, Argentina. Instituto de Historia, Resistencia

Kanter H (1932) La cuenca cerrada de la Mar Chiquita en el norte de la Argentina. Bol Acad Nac Cienc Córdoba (Argentina) 22:285–232

Kröhling D, Iriondo M (1999) Upper quaternary palaeoclimates of the Mar Chiquita area, North Pampa, Argentina. Quat Int 57/58:149–163

Kruck W, Helm F, Geyh M, Suriano J, Marengo H, Pereyra F (2011) Late pleistocene-holocene history of Chaco-Pampa sediments in Argentina and Paraguay. Quatern Sci J 60:188–202

Lewis JC (1993) Foods and feeding ecology. In: Baskett TS, Sayre MW, Tomlinson RE, Mirarchi RE (eds) Ecology and management of the mourning dove. Stackpole Books, Harrisburg, pp 181–204

Martinez DE (1991) Caracterización geoquímica de las aguas de la Laguna Mar Chiquita, provin- cia de Córdoba. Universidad Nacional de Córdoba, Córdoba

Martinez DE (1995) Changes in the ionic composition of a saline lake, Mar Chiquita, province of Cordoba, Argentina. Int J Salt Lake Res 4(4):25–44

McGlue MM, Ellis GS, Cohen AS (2015) Modern muds of Laguna Mar Chiquita (Argentina): Particle size and organic matter geochemical trends from a large saline lake in the thick-skinned Andean foreland. In: Larsen D, Egenhoff SO, Fishman NS (eds) Paying attention to mudrocks: priceless!: 1–18. Geological Society of America Special Paper

Mon R, Gutierrez A (2009) The Mar Chiquita Lake: an indicator of intraplate deformation in the central plain of Argentina. Geomorphology 111:111–122

Ninci C (1919) Los Bañados de 'El Tío. In: Fuchs G (ed) Memoria del Ministerio de Obras Públcas de Córdoba. Mayo de 1928–Mayo de 1929. Ministerio de Obras Públicas de la Provincia de Córdoba, Córdoba, pp 11–33

Piovano E, Larizzatti F, Favarol D, Oliveira S, Damatto S, Mazilli B, Aristegui D (2004a) Geochemical response of a closed-lake basin to 20th century recurring droughts/wet intervals in the subtropical Pampean Plains of South America. J Limnol 63:21–32

Piovano EL, Ariztegui D, Bernasconi SM, McKenzie JA (2004b) Stable isotopic record of hydro-logical changes in subtropical Laguna Mar Chiquita (Argentina) over the last 230 years. Holocene 14:525–535

Piovano EL, Zanori GA, Ariztegui D (2006) Historia geologica y registro humano. In: Bucher EH (ed) Bañados del rio Dulce y Laguna Mar Chiquita (Córdoba, Argentina). Academia Nacional de Ciencias (Córdoba, Argentina), Córdoba

Piovano E, Ariztegui D, Córdoba F, Cioccale M, Sylvestre F (2009) Hydrological variability in South America below the tropic of Capricorn (Pampas and Patagonia, Argentina) during the last 13.0 Ka. In: Vimeux F, Sylvestre F, Khodri M (eds) Past climate variability in South America and surrounding regions. Springer, Berlin, pp 323–351

Roverano AA (1955) El Río Salado en la historia. Andres A. Roverano, Santa Fe

Soldano FA (1947) Régimen y aprovechamiento de la red fluvial argentina. Parte II. Ríos de la región árida y de la meseta patagónica. Buenos Aires.: Editorial Cimera

Chapter 3
Hydrology and Climate

3.1 Introduction

As a basic principle in wetland ecology, any modification to the hydrologic regime, whether at a site or watershed scale, will result in changes in a wetland. In this sense, (Maltby and Barker 2009) consider that "hydrology is the single most important determinant of the understanding of wetlands and wetland processes, and a key tool for wetland management." Therefore, before any management decision is made, it is essential to thoroughly understand the role and characteristics of the specific wetland hydrology. In the case of Mar Chiquita, there is a multitude of different factors, both natural and human-induced, that can lead to a change in its hydrologic regime. Among them, factors of primary concern are climate change, water requirements for human use, and ecosystem and biodiversity conservation (Bucher et al. 2006). In order to fully understand the hydrological characteristics and dynamics of the Mar Chiquita wetland, a broad, integrative approach is required, which includes the climate that controls rainfall and evapotranspiration, the catchment basin and the tributary rivers, and the water balance in the Mar Chiquita Lake and the Rio Dulce wetland.

3.2 Climate

The climate in the Mar Chiquita basin is subtropical, semiarid, and characterized by a typical monsoon pattern, with warm and wet summers (October–March) and dry and cold winters (April–September). Given the size of the basin area, climatic conditions change according to locations, following two main directional gradients. Temperature decreases from north to south, with the annual average ranging from 23 °C in the north to 17 °C in the southern extreme. Annual rainfall decreases from about 1000 mm in the east to less than 700 mm in the west up to a close distance

© Springer Nature Switzerland AG 2019
E. H. Bucher, *The Mar Chiquita Salt Lake (Córdoba, Argentina)*,
https://doi.org/10.1007/978-3-030-15812-5_3

Fig. 3.1 Drainage basin of
Mar Chiquita. (1)
Aconquija mountains, (2)
Salado River, (3) Rio
Hondo dam, (4) Dulce
River, (5) Ambargasta salt
flats, (6) Saladillo branch
of the Dulce River, (7)
Western dispersed creeks,
(8) Rio Dulce wetland, (9)
Sierras de Córdoba
mountains, (10) Primero
River, (11) Mar Chiquita
Lake, (12) Segundo River.
Dotted lines indicate
provincial limits

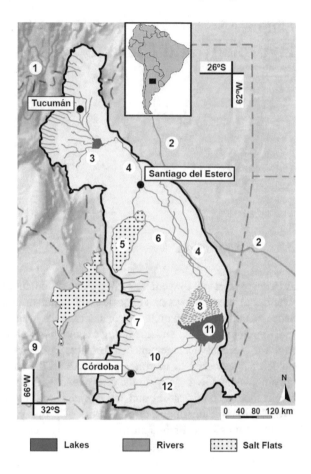

from the western Aconquija mountains, where it increases again due to the orographic effect. Given the elevation of the Aconquija range (over 5000 m) (Fig. 3.1)
and the dominance of easterly winds that bring vapor from the Atlantic Ocean, the
orographic effect is substantial, and rainfall increases to over 2500 mm on the eastern slopes. High rainfall, together with persistent fog banks during winter, allows
the development of a dense tropical cloud forest, known locally as *Yunga*. Rainwater
is collected almost entirely by a hydrographic network that flows downward into the
Dulce River (named Sali River in the Tucuman province).

A similar situation, although at a much smaller scale, takes place in the southern
portion of the Mar Chiquita basin, located to the south in the province of Córdoba.
There, the two remaining main tributaries, the Primero and Segundo rivers, originate in the Sierras de Córdoba range, with a maximum elevation of 3000 m
(Fig. 3.1).

Recent Changes in the Regional Climate A remarkable increase in precipitation occurred over most of subtropical Argentina since 1970, ranging between 10% and 30% according to regions (Perez et al. 2015). The entire Mar Chiquita basin was subjected to this phenomenon, as shown by the displacement of isohyets in the region (Fig. 3.2). This huge rainfall increase implied a substantial additional amount of water flowing to Mar Chiquita, besides the extra rainfall falling on the lake surface (Troin et al. 2010), all of which resulted in the marked increase of water level recorded in the Mar Chiquita Lake. Incidentally, the westward displacement of the 800 mm isohyet (which indicates the lower limit for nonirrigated agriculture) allowed the expansion of the area under cultivation in about 100 km to the west in northern Argentina (Barros et al. 2015) (Fig. 3.2).

Fig. 3.2 Changes in annual precipitation during the 1930–2000 period in the Mar Chiquita drainage basin. (**A**) Decadal average variations. (**B**) Displacement of the isohyets in the region occupied by the Mar Chiquita basin during this period. Dotted lines indicate provincial boundaries. (From Bucher 2006)

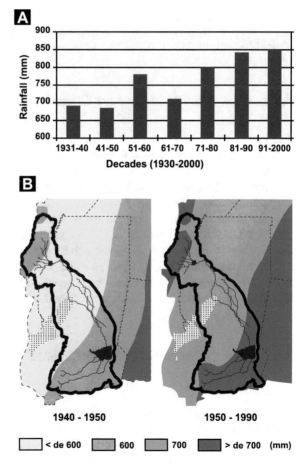

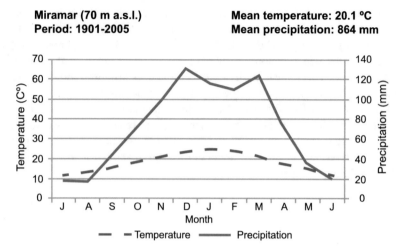

Fig. 3.3 Climograph of Miramar (Mar Chiquita). According to Walter and Lieth (1967), in the months when the precipitation line is above temperature, the water balance is positive, whereas the reverse occurs when the temperature is above precipitation

3.2.1   The Climate in the Mar Chiquita Area

The local climate is summarized in Fig. 3.3. Mean annual temperature is 19 °C, mean maximum of the warmest month (January) is 32 °C, and mean minimal of the coldest month (July) is 3.5 °C. Mean annual rainfall ranges between 800 and 900 mm, concentrated in the summer months (October to March). The annual evapotranspiration rate (the only significant water loss in the system) has been estimated at about 1300 mm (Troin et al. 2010). Water evaporation is favored by the fact that Mar Chiquita is located in the open Pampas plains, where strong winds are frequent and intense on occasions, and also by the pan-like shape of the lake, favoring water evaporation, homogenization of the water column, and sediment removal and mixing. During the warm season, east and northeast winds are predominant, whereas in winter, dominant winds blow from the south and southwest.

3.2.2   The Influence of the Mar Chiquita Wetland on the Local Climate

The whole Mar Chiquita wetland (including the lake and the Dulce River wetland) is the primary pathway of water loss through evapotranspiration in this closed basin (Rodriguez et al. 2006; Troin et al. 2010). It is very likely that the wetland may have a significant effect on the local climate, considering its size of about 10,000 km².

More research on this topic will be necessary from several points of view, including the assessment of the environmental services provided by the lake. According to Jørgensen (2010), it is expected that Mar Chiquita may reduce the maximum temperature and increase minimum temperature all year round and also increase precipitation over the lake area and in the nearby region.

3.3 Hydrology

3.3.1 The Drainage Basin

The Mar Chiquita Lake is the depocenter of an endorheic basin about 127,000 km^2 included between 26–32°S and 62–66°W, shared by the provinces of Córdoba, Santiago del Estero, Tucuman, and Salta (Fig. 3.1). This large basin is part of the Chaco-Pampa Plain, a geological region that extends from the central and northern lowlands of Argentina to the Paraguayan and Bolivian Chaco regions (Chap. 2).

Three major river systems are included in this basin: Primero (also known as Suquia), Segundo (also known as Xanaes), and Dulce. There are also several small creeks on the western coast of the lake, known as Western dispersed creeks. Groundwater is another potential factor of both inflow and outflow from the lake that needs to be considered, even though it appears of limited significance (Stappenbeck 1926) (Fig. 3.1).

These three main rivers have a strong seasonal regime, with streamflow reaching peak values in summer and autumn, following the monsoon type of the regional rainfall pattern. All of them have dams regulating their flow. Their main morphological characteristics are summarized in Table 3.1 and Fig. 3.1.

Table 3.1 Main hydrological characteristics of the Mar Chiquita tributary rivers

Rivers	Drainage basin	Length	Streamflow	
	km^2	Km	m^3/s	hm^3/year
Dulce (highstands)	72,000	812	83	2619
Dulce (lowstands)			137	4339
Primero	7500	220	10	366
Segundo	6700	340	9.8	359
Total (lowstand period)	86,200		102.8	3344
Total (highstand period)			156.8	5064

3.3.2   The Rivers

The Dulce River It is the main tributary of Mar Chiquita, which on average provides about 80% of the total water volume reaching the lake (Table 3.1). The river headwaters are located in the limit between the provinces of Salta and Tucumán (NW region of Argentina) on the Sierras Calchaquies and Nevados de Aconquija mountains, at height of about 5000 m a.s.l. The river has two main dams along its entire course: Escaba in the upper course, and Rio Hondo, in the middle river. At least two more dams are projected for the lower Dulce River (Gallego 2012). The Rio Hondo dam holds about 200,000 ha of irrigated land in Santiago del Estero province. The annual average discharge from the Rio Hondo dam varied markedly during the last 90 years. During the lowstand 1926–1972 period, discharge was much lower than during the highstand 1972–2010 period (Table 3.1) (Gallego 2012). About 100 km downstream of the Rio Hondo dam, the Dulce River opens in a small branch (known as Saladillo River), which flows southward through the Ambargasta salt pan, increasing the salinity of its waters, and ends joining the main Dulce River stream again (Fig. 3.1). From this connection point onward, the river flows through a vast terminal floodplain wetland locally known as Bañados del Rio Dulce. This area has a minimal low slope that allows expansion of waters on extensive wetlands and a wandering course of the main river, forming a complex braided pattern (Chap. 10). During the lowstand periods, several lagoons are visible in the Dulce River wetlands, which become submerged and integrated with the main lake during the highstands. Finally, it opens into the Mar Chiquita Lake through a wide delta.

Primero River It is also known as Suquia River. This river originates from the San Roque dam, on the Sierras de Cordoba hills (about 2000 m a.s.l.), and flows into Mar Chiquita through the Del Plata lagoon (Table 3.1, Fig. 3.1.) There is one dam along its course (San Roque) (Wunderlin 2018).

Segundo River It is also known as Xanaes River, which flows from the Los Molinos dam on the Sierras de Cordoba hills (about 2000 m a.s.l.). This river flows into the Mar Chiquita Lake mainly through the Plujunta River (to which was diverted from the Segundo River in 1927 (Fig. 3.1). Occasionally this river overflows through the old riverbed known as Rio Segundo Viejo (Chap. 2). There is one dam along its course (Los Molinos).

Western Dispersed Creeks Toward the west of the Mar Chiquita Lake, there are several small creeks that flow from the Sierras de Cordoba to Mar Chiquita. They usually infiltrate into the soil before reaching the lake except after exceptional rains (Cioccale 1999) (Fig. 3.1).

3.3.3   The Lake

3.3.3.1   Lake Morphology, Volume, and Area

The lake is a very shallow water body, with an absolute maximum depth of about 10 m at the highest recorded water level. The bottom has a slight north-south slope, with the deepest area located close to the southern coast (Fig. 3.4). An important discontinuity point appears at the 67 m a.s.l. level, which corresponds to the highest lake level before the substantial expansion that started in the 1970s. Above this level, the elevation gradient is minimal, as shown by the vast extension of the area included between this level and the maximum extension of the lake in highstand periods (Fig. 3.4). Given the shallowness of this recently flooded area (slope <1%), the coastline may expand and retreat long distances even with relatively small variations in water level, which has resulted in the development of a large salt flat areas, from where massive salt dust storms may be generated under favorable conditions

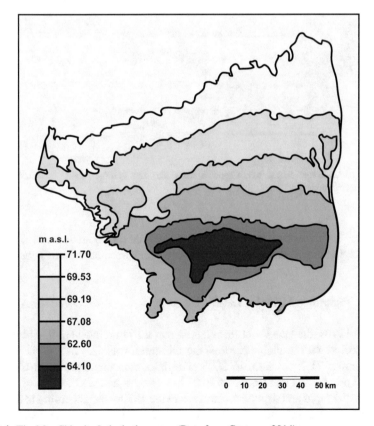

Fig. 3.4 The Mar Chiquita Lake bathymetry. (Data from Costanza 2014)

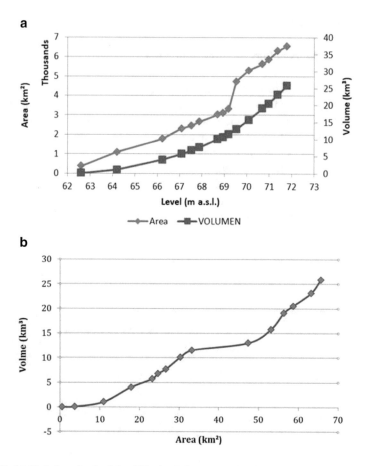

Fig. 3.5 (**a**) Variations in the Mar Chiquita Lake area and volume related to water level, (**b**) volume-area relationship

(see Chap. 2). These particular characteristics of the lake bottom are also reflected in the correlation pattern between the lake level and area and volume (Fig. 3.5).

3.3.3.2   Variations in Water Level

Since the 1970s, the lake water level has shown a 10 m range variation along time. For the period with available records, the minimum was 62.4 m a.s.l. in 1972 and the maximum 71.79 m a.s.l. in 2003. The lake area increased from 63 km² to 6553 km² and the volume from 0.08 to 25 km³ (Rodriguez et al. 2006).

Before the rapid and significant increase in the 1970s, and according to historical data (Piovano et al. 2002) and indirect data based on salinity measurements (Bucher and Bucher 2006), the lake level peaked in 1890 at about 67.5 m a.s.l. and since then

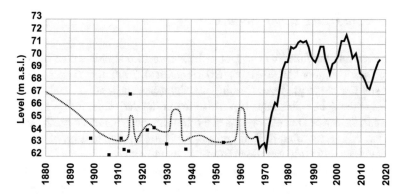

Fig. 3.6 Variations in the Mar Chiquita Lake water level along the 1890–2015 period. Data from 1890 to 1967 period (dotted line) are based on local historical information according to Piovano et al. (2002). Data from 1967 onward (solid line) correspond to instrumental records, Data for this period from (Bucher and Bucher 2006; Curletto 2014), and records collected by the Mar Chiquita Biological Station (IDEA CONICET and Universidad Nacional de C'ordoba). Isolated small dots represent water level estimated from salinity samples

decreased and remained oscillating between 64 and 65 m a.s.l. During this period there were short-term peaks in the 1910, 1930, and 1960 decades, always below the 67 m a.s.l. discontinuity limit mentioned previously (Fig. 3.6).

3.3.3.3 Groundwater

There is limited information on the significance of groundwater in the water balance of the Mar Chiquita Lake. Concerning water loss due to groundwater seepage, both studies by Troin et al. (2010) using chloride balance and DaPeña and Panarello (2001) based on isotopic studies conclude that seepage is not significant and therefore has no influence on the water balance of the lake.

Regarding water recharge of the lake by aquifers, the general flow direction of shallow aquifers around Mar Chiquita converges to the lake, becoming more intense during high rain periods (Blarasin et al. 2014). However, according to Stappenbeck (1926), the contribution of groundwater to the Mar Chiquita water balance is expected to be limited, since predominant soils in the area are fine textured and with low permeability. With regard to deeper, confined aquifers, they are isolated from the lake, except in the case of artesian wells that are drilled and abandoned around the lake (Blarasin et al. 2014). Even if their number is limited, they remain discharging water into the lake for decades, becoming a waste of fresh water that deserves concern.

With regard to outflows, given that Mar Chiquita is a closed basin with no river outlet, losses are restricted to evapotranspiration (sum of evaporation plus plant transpiration) and underground losses (seepage), besides human use.

3.3.4   The Dulce River Wetland

Before reaching the lake, the Rio Dulce feeds a vast floodplain called the Bañados del Rio Dulce (of about 10,000 km²) (Rodriguez et al. 2006) that is characterized by a very shallow slope gradient, forming marshy grasslands, ponds, and small lakes (Fig. 3.1) (see details in Chap. 10). This area is flooded annually by the seasonal increase in the River Dulce flow caused by the summer rains in the catchment basin (see Sect. 3.2). According to Rodriguez et al. (2006), the extension of the flooded area is subjected to wide variations from 325 to 3400 km².

However, no clear link between surface variations and Rio Dulce discharge or lake level was determined, probably due to the difficulty in assessing evapotranspiration in this vast area that lacks an adequate network of measuring stations. According to Troin et al. (2010), evapotranspiration in this wetland is one of the main sources of water loss in the whole Mar Chiquita system (Troin et al. 2010), since the lake has no surface outlet and the only significant water loss is through evaporation.

With regard to groundwater, in this subregion, the aquifer is contained in fine to very fine sediments at very shallow depth (1–2 m), with high salt content. Recharge is associated with both rainfall and Dulce River annual flooding. Soil sediments may act as a buffer factor, absorbing water and attenuating floodings, a difficult to assess factor when modeling the water balance of the Mar Chiquita hydrological system (Blarasin et al. 2014) (Fig. 3.7).

3.3.5   Hydrological Models

Mathematical modeling of watershed hydrology is an essential tool for understanding and managing the water balance of a wetland and is employed to address a broad spectrum of environmental and water resource problems. Although the hydric balance concept is simple, constructing mathematical models that accurately reflect that balance is not an easy task, given the size and complexity of such a vast hydrological system as Mar Chiquita wetlands. Likewise, developing models of improved predictive power depends on the availability and quality of the database.

Significant advances have been made in modeling the Mar Chiquita hydrological system, although much has still to be made. Here we describe two recently developed models, their results, and the potential applications in the Mar Chiquita wetland system. They are of great help in main management questions, including (a) controlling the lake water level and (b) managing the annual sheet-flood pulse in the Dulce River wetlands. No doubt, models are an essential tool for the sustainable management of the wetland (Wurtsbaugh et al. 2017).

Fig. 3.7 Flood expanding from the Dulce River into the Rio Dulce wetland. (From Bucher 2006)

3.3.5.1 Lambda Model

This is a water balance model being developed by researchers of Laboratorio de Hidráulica y Centro de Estudios y Tecnología del Agua (CETA) CONICET-Universidad Nacional de Córdoba (Argentina), at present in its version 2, and further versions are being developed (Curletto 2014). The model inputs include the following parameters: streamflow of the tributary rivers, rainfall in the Mar Chiquita Lake and the Dulce River wetlands, and evaporation from the lake and the Dulce River wetlands. Groundwater influx and seepage were not measured because (a) adequate field data were not available and (b) existing published evidence indicates that these two factors do not appear significant (Stappenbeck 1926; DaPeña and Panarello 2001).

Lambda has a spatially semi-distributed structure, separating three geographical subsystems: medium Dulce River, Dulce River wetlands, and the Mar Chiquita Lake. In addition, the model includes a complementary section dealing with the dynamics of the annual flood pulse in the Dulce River wetland.

The Lambda model has produced results of direct application to critical water management issues in the Mar Chiquita wetland, including the following:

1. Inflow required from the Dulce river to restore and sustain the lake water level and salinity (Rodriguez et al. 2006). Results are shown summarized in Table 3.2).

Table 3.2 Water flows
from the Dulce River
required to maintain different
water levels of the Mar
Chiquita Lake. From
Rodriguez et al. (2006)

Water level	Salinity	Flow required
m a.s.l.	(g/L)	(m³/s)
71,5	25	142,2
71,0	30	105,7
70,5	34	89,0
70,0	40	80,3
69,5	45	74,1
69,0	52	67,8
68,5	60	58,3
68,0	69	49,7
67,5	79	40,3
67,0	91	30,7
66,5	104	21,9
66,0	120	14,6
65,5	137	8,9
65,0	156	4,6
64,5	181	0,1

2. The inflow pulse needed for ensuring the annual flooding of the Rio Dulce wetland. The Dulce River flooding submodel estimates the amount of water required
   to be released by the Rio Hondo dam for three consecutive months to produce a
   sheet flood of a given size in the Dulce River wetlands. Initial results showed that
   low volume releases up to 250 hm³ floods an area of less than 500 km², for
   medium volume releases (250–2000 hm³) the area flooded is between 2000 and
   2800 km², and releases of over 2800 km² may cause floods of 3500 km² or even
   larger (Rodriguez et al. 2006; Curletto 2014). These values require further refine-
   ment, considering that new observations suggest that the observed relationship
   may be influenced by the lake level (that provides the base level for the flooding
   waters) and by the degree of underground water saturation in the Dulce River
   wetlands (Curletto 2014). In addition, analysis of satellite images showed that
   the displacement of the sheet flood along the Rio Dulce wetland takes about
   4 weeks (Curletto 2014).
3. A risk analysis of the maximum water levels to be expected by the coastal town
   of Miramar (including maximum storm level resulting from wind and waves
   generated by the storm) was produced by Pagot et al. (2014). This study required
   producing a new bathymetry of the lake which was rebuilt using remote sensing
   techniques.

3.3.5.2 Mar Chiquita Hydrological Model of a Closed Lake

This model was developed by researchers of CEREGE, Aix-Marseille Université,
CNRS, IRD (France), and CICTERRA-CIGeS, Universidad Nacional de Córdoba,
Argentina (Troin et al. 2010). The model includes quantification of the lake water

balance and the distribution of lake water inflows between local influences (i.e., direct precipitation) and long-distance influences (i.e., catchment contribution) mainly due to upstream river flows.

The study results indicate that the catchment contributes, on average, 33% to total inflows during the simulation period (1967–2006), and this proportion was highly variable, from a zero value (e.g., dry years of 1972 or 1989) to almost 71% in wet years. It also showed that 92% of lake level variations were attributed to a runoff increase in the upper Dulce River catchment, suggesting a tropical climatic influence.

Evaporative flux was identified as the most sensitive parameter in the water balance of the system. Moreover, a strongly negative water balance in the ungauged part of the catchment (particularly the Dulce River wetlands) could be attributed to evapotranspiration. With regard the possibility of groundwater seepage or inflows, the chloride balance indicated that the lake is hydrologically closed, supporting similar conclusions from other studies in the region (DaPeña and Panarello 2001).

With regard to the future of modeling of the Mar Chiquita hydrological system, authors of the two models described previously mention that the main challenge for model development and implementation in Mar Chiquita comes from insufficient, discontinuous, and sparsely distributed data collecting points. For example, ungauged downstream surfaces at the time of these studies represented approximately 80% of the catchment. Clearly, strengthening the data-gathering system in the area appears as a priority for improving and widening the use of hydrological modeling in the region.

References

Barros VR, Boninsegna JA, Camilloni IA, Chidiak M, Magrín GO, Rusticucci M (2015) Climate change in Argentina: trends, projections, impacts and adaptation. WIREs Clim Chang 2:151–169

Blarasin M, Cabrera A, Matteola E (2014) Aguas subterráneas de la provincia de Córdoba. UniRío Editora E-Book, Rio Cuarto

Bucher EH (ed) (2006) Bañados del Rio Dulce y Laguna Mar Chiquita, Córdoba, Argentina. Academia Nacional de Ciencias, Córdoba

Bucher E, Bucher AE (2006) Limnología Física y Química. In: Bucher EH (ed) Bañados del Rio Dulce y Laguna Mar Chiquita (Córdoba, Argentina). Academia Nacional de Ciencias, Córdoba, pp 79–101

Bucher E, Coria R, Curto E, Lima J (2006) Conservación y uso sustentable. In: Bucher E (ed) Bañados del Rio Dulce y Laguna Mar Chiquita (Córdoba, Argentina). Academia Nacional de Ciencias (Córdoba, Argentina), Córdoba, pp 327–341

Cioccale M (1999) Climatic fluctuation in the Central region of Argentina in the last 1000 years. Quat Int 62:35–47

Curletto LM (2014) Análisis de datos hidrometeorológicos aplicados al Balance Hidrol'ogico de los bañados del rio Dulce y Laguna Mar Chiquita. In: Centro de Estudios y Tecnología del Agua. Córdoba, Argentina: Universidad Nacional de Córdoba

DaPeña C, Panarello HO (2001) Isotopic study of the "Laguna Mar Chiquita", Córdoba, Argentina and its hydrogeological and paleoclimatological implications. IAEA – TEC-DOC- Series1206: 7-15. ISSN 1011-4289

Gallego A (2012) Santiago del Estero y el agua: crónica de una relación controvertida Santiago del Estero. Lucrecia Editorial, Argentina

Jørgensen SE (2010) A review of recent developments in lake modeling. Ecol Model 22(4):689–692

Maltby E, Barker T (2009) The wetlands handbook. Wiley-Blackwell, Chichester

Pagot M, Hillman G, Pozzi Piacenza C, Gyssels P, Patalano A, Rodriguez A (2014) Maximum water level in Mar Chiquita, Lagoon, Cordoba, Argentina. Tecnol Cienc Agua 5:115–128

Perez S, Serra E, Momo F, Massobrio M (2015) Changes in average annual precipitation in Argentina's Pampa region and their possible causes. Climate 3:157–164

Piovano EL, Ariztegui D Damatto Moreiras S (2002) Recent environmental changes in Laguna Mar Chiquita (central Argentina): a sedimentary model for a highly variable saline lake. Sedimentology 49. https://doi.org/10.1046/j.1365-3091.2002.00503.x

Rodriguez A, Pagot M, Hillman G, Pozzi C, Plencovich G, Caamaño Nelli G, Curto E, Bucher E (2006) Modelo de Simulación Hidrológica Bañados del Rio Dulce y Laguna Mar Chiquita (Córdoba, Argentina). In: Bucher E (ed) Bañados del Rio Dulce y Laguna Mar Chiquita (Córdoba, Argentina). Academia Nacional de Ciencias, Córdoba, Argentina, Córdoba

Stappenbeck R (1926) Geología y aguas subterráneas de la Pampa. Editorial Pangea, Córdoba

Troin M, Vallet C, Suylvestre F, Piovano E (2010) Hydrological modeling of a closed lake (Laguna Mar Chiquita, Argentina) in the context of 20th-century climatic changes. J Hydrol 393:233–244

Walter H, Lieth H (1967) Klimadiagram-Weltatlas. Gustav Fischer Verlag, Jena

Wunderlin DA (2018) The Suquía River Basin (Córdoba, Argentina). An integrated study on its hydrology, pollution, effects on native biota and models to evaluate changes in water quality. Springer

Wurtsbaugh WA, Miller C, Null SE, DeRose R, Justin PW, Hahnenberger M, Howe F, Moore J (2017) Decline of the world's saline lakes. Nat Geosci 10:816–821

Chapter 4
Limnology

4.1 Introduction

Limnological research in Mar Chiquita has been limited until recently, with the exception of a pioneer, high-quality monograph published in 1945 by the Argentine Academy of Sciences (Seckt 1945). Unfortunately, Seckt's research was published in Spanish and remained unknown by the international community. More recently, Bucher and Bucher (2006) and Bucher and Abril (2006) provided an updated summary of the information available on the lake's limnology. Pilati et al. (2016) produced a 2-year study of the annual limnological cycle of the lake, and Bucher and Stein (2016) reported the generation of massive salt dust storms in the lake. The present review is structured around comparative analysis of the limnological characteristics and dynamics of the lake between the lowstand period of low water level and high salinity and the recent highstand situation of high water level and lower salinity (see Chaps. 3 and 12). This review attempts at providing an updated summary of the existing information and also identifying knowledge gaps and research needs in terms of both basic and applied scientific priorities.

4.2 Physical Limnology

Water Transparency Based on the limited available information, this parameter ranges between 0.24 and 0.72 m (Bucher and Bucher 2006; Pilati et al. 2016). Exceptionally high values were reported in 1989 when water transparency at the center of the lake ranged between 0.8 and 1.5 m (water salinity = 36 g/L) (Reati et al. 1997).

© Springer Nature Switzerland AG 2019
E. H. Bucher, *The Mar Chiquita Salt Lake (Córdoba, Argentina)*,
https://doi.org/10.1007/978-3-030-15812-5_4

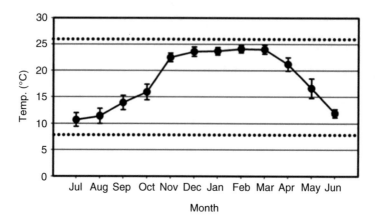

Fig. 4.1 Mar Chiquita annual variation in the mean and standard error of monthly temperature (°C), recorded at 1.5 m below water surface during the 2008–2009 period. From Bucher (2006). Dotted lines indicate the absolute minimum and maximum temperature recorded during the whole period. (Data from the Mar Chiquita Biological Station, Universidad Nacional de Córdoba)

Water Temperature According to Pilati et al. (2016), water temperature in Mar Chiquita in 2010–2011 (measured at 1.5 m deep) followed a well-defined seasonal pattern, with a mild winter and a warm summer (Fig. 4.1). By early spring (September), water temperature steadily rose and then remained above 20 °C for 6 months. Winter (July) mean temperature was 10.7 °C, while in summer (February) reached 24.1 °C. Non-systematic temperature profiles indicated uniform values from the surface down to 4 m in all seasons except in early spring, when bottom temperatures 2.5 °C lower than at the surface were recorded. In summary, the available evidence indicates that Mar Chiquita is polymictic, as it would be expected given its shallowness, although some degree of thermal stratification may be expected during the warm season.

Water Color This parameter may vary considerably in Mar Chiquita Lake. The usual tone is yellowish green, which may sometimes turn to a more intense green during high biological productivity periods (Bucher and Abril 2006; Bucher and Bucher 2006) and may occasionally turn red in small ponds close to the lake coast, due to microbial blooms. According to Oren (2002), the available evidence indicates that bacterioruberin and other carotenoid pigments of the Halobacteriaceae family are generally responsible for most of the red color of the saline waters (Fig. 4.2).

Dissolved Oxygen The dissolved oxygen concentration in water is considered one of the most critical parameters in limnology and particularly in salt lakes, since oxygen solubility decreases with increasing water salinity. According to Hammer (1986), in salt lakes, oxygen level is frequently oversaturated on the surface but drops rapidly with depth, particularly in warm days. This tendency was confirmed

Fig. 4.2 Red water color in the Mar Chiquita lake waters, generated by bacterial pigments in high salinity waters. (From Bucher 2006)

by data from Mar Chiquita. In a year-long survey, Pilati et al. (2016) found that in winter values were similar along the entire water column, whereas in spring a moderate decrease with depth was detected, which became more marked in summer. At this time oxygen reached a minimal bottom concentration of 2.2 mg/L, and the sampled material was a black mud with a strong sulfide smell, indicating anoxic condition. A similar tendency for lower oxygen values in deeper samples along a 3-year sampling was reported by Martinez (1991).

Foam Foam generation is a characteristic feature of the salt lakes that differentiate them from freshwater lakes, being frequent in the Great Salt Lake in Utah, USA (Wallace Gwynn 2004). In Mar Chiquita foam generation is frequent, and it appears in different forms: shore accumulations, Langmuir circulation streaks, and winding lines (Fig. 4.3).

Shore foam accumulations, sometimes of over half meter thick, occur especially during and after strong winds that bring high waves to the shores. Langmuir streaks consist of long lines of foam parallel to the wind direction across the water surface, called "windrows," which result from complex horizontal and vertical movements of the water bubbles (Thorpe 2004). Common in oceans, the Langmuir streaks are frequently seen in Mar Chiquita on windy days (Fig. 4.3). Winding lines are frequently seen in Mar Chiquita close to the mouth of tributary rivers, notably the Segundo River, resulting from the contact between the river colder fresh water and the saltier and denser lake water (Fig. 4.3). Foam generation in salt lakes is a

Fig. 4.3 Foam in the Mar Chiquita Lake. Left, parallel Langmuir circulation streaks and also a single, broader winding line. Right, detail of a winding line with bubbles indicating intense biological activity

complex phenomenon not yet been completely clarified. The more accepted theory proposes that foam in saline lakes is caused by the presence of surfactants naturally produced by blue-green algae species (Cyanophyceae) (Wallace Gwynn 2002). These organic surfactants are part of a large variety of plant material that when dissolved in water is referred to as dissolved organic carbon (DOC).

4.3 Chemical Limnology

4.3.1 Water

In Mar Chiquita, salinity has ranged between an extreme hypersaline condition, with maximum values of 360 g/L during the lowstand period, and a mesosaline condition during the recent highstand period, with minimum values of 25 g/L in response to variations in the lake water volume.

The main characteristic of the water of Mar Chiquita Lake is the strong dominance of the anions chloride and sulfate and of the cations sodium, calcium, and magnesium, as evidenced in the following percent values obtained in 1978 at a salinity level of 79 g/L: sodium chloride (halite) ($NaCl$), 78%; sodium sulfate (mirabilite or Glauber's salt) (Na_2SO_4), 17%; calcium sulfate (gypsum) ($CaSO_4$), 2.3%; and magnesium sulfate (Epsom salt) ($MgSO_4$, 2.0%). The relationship between water level and salinity is shown in Fig. 4.4.

The chemical composition of Mar Chiquita water is similar to that of seawater but differs mainly in a much higher sulfur content (Table 4.1), higher also than the Great Salt Lake of Utah. This combination of salts is one of the most common among salt lakes worldwide, particularly in Argentina, which results from the geological substrate of the tributaries riverbed (Hammer 1986).

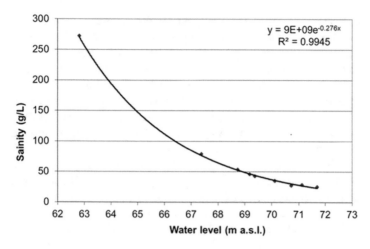

Fig. 4.4 The relationship between the Mar Chiquita Lake water level and salinity. The equation on the top right indicates its mathematical expression. (From Bucher 2006)

Table 4.1 Chemical composition of sediments of the Mar Chiquita Lake. From Bucher and Bucher (2006)

Parameter	Year		
	1978	1994	2002
Ph	6.95–7.55	–	7.48–8.41
Silica (%)	–	36.2–63.4	–
Aluminum (%)	–	9.11–13.2	–
Iron (%)	0.91–2.37	1.59–5.65	3.4–5.38
Manganese (%)	0.03–0.71	0.05–0.15	–
Mg (%)	0.73–1.79	1.62–6.78	–
Calcium (%)	2.43–8.75	2.26–12.7	–
Sodium (%)	3.07–6.92	0.34–1.73	–
Potassium (%)	0.08–1.02	1.52–2.67	–
Titanium (%)	–	0.61–1.00	–
Phosphorus (%)	–	0.22–0.39	0.02–0.38
Carbon (%)	–	0.06–8.43	–
Sulfur (%)	–	0.14–8.28	–
Lithium (%)	0.06–0.18	–	–
Chlorides (%)	1.66–4.66	–	–
Sulfates (%)	0.62–3.01	–	–
Calcium carbonate (%)	2.61–4.14	–	–
Sulfur (%)	0.004–0.076	–	–
Organic matter (%)	4.1–18.0*	5.69–22.9	0.46–16.32
Fulvic acids (%)	–	–	0.26
Humic acids (%)	–	–	0.55
Total nitrogen (%)	–	–	0.18–0.60

The relative proportion of each compound in the lake water changes as a function of its solubility. For example, the chloride/sulfate proportion increases with total salt concentration in water, since sulfates are less soluble than chlorides and therefore reach the saturation level at a lower concentration. Over the saturation level, the excess of sulfate precipitates in the sediment, while the chloride concentration continues increasing. For example, in Mar Chiquita, the proportion of dissolved chlorides in water was three times higher when water salinity was between 28 and 79 g/L, whereas at a salinity of 280 g/L, chlorides were five times more abundant than sulfates (data from Martinez 1991). In addition, the solubility of sulfates decreases even more with decreasing water temperature, whereas solubility of chlorides remains almost constant. This property explains that, during lowstand periods and on cold winter days, a considerable accumulation of sodium sulfate crystals appears on the shore, locally known as "winter salt." These salt accumulations are particularly prone to be transported by wind and generate salt dust storms (Bucher and Stein 2016) (see also Chap. 3).

4.3.2 Sediments

The Mar Chiquita lake-floor sediment is heterogeneous, since it varies within a wide range regarding texture (sandy, silty, clay) and consistency, from compact to very soft, such as black mud. These differences are related to deposition in river deltas, areas, variations in the lake level and expansion on coastal areas, and the existence of long-term depositional profound areas (Chap. 3).

The mean values of physical and chemical characteristics of lake sediments taken along time are indicated in Table 4.2, in which samples differing in water salinity level are compared. On average, the dominant mineral components include: clay minerals, mostly abundant mica-like illite, with lesser amounts of calcium-rich smectite and minor to trace amounts of kaolinite. Calcite is the dominant carbonate (McGlue et al. 2015). According to Martinez (1991), the relative proportion of the main elements found in the sediments (in weight) is as follows: silica 44.7%, aluminum 10.7%, calcium 7.8%, carbon 4.5%, manganese 4.0%, magnesium 3.9%, sulfur 3.4%, iron 2.5%, potassium 2.1%, and sodium 0.7%.

The authigenic, organic-rich black mud is an important component of the Mar Chiquita sediments, not only in terms of its own nature but also because of its value as a health product attracts tourism to the region. Organic black mud is a common feature in saline lakes; in most of them, it is also considered of therapeutic value, particularly in the Dead Sea (Hammer 1986). The chemical composition of the black mud in Mar Chiquita includes the following mineral components: ~33% clay minerals, 10% calcite, 7% quartz, and 4.5% halite plus gypsum, with the remaining portion made up of feldspars, altered volcanic glass, and amorphous material (organic matter and diatoms) (McGlue et al. 2015).

Table 4.2 Chemical composition of sediments of the Mar Chiquita Lake obtained with different salinity laves. From Bucher and Bucher (2006)

	Salinity				
	High	Medium	Low	Very low	
Parameter	1970	1977	1989	1986	2002
Density	–	1056	1022	–	–
Dissolved solids (g/L)	283	79	35	28	27
Ph	9.45	8	8.53	8.3	8
Sodium (g/L)	107.73	28.2	12.52	9.96	–
Potassium (g/L)	1.56	0.28	0.14	0.13	–
Calcium (g/L)	1.02	0.53	0.35	0.28	–
Magnesium (g/L)	0.72	0.36	0.25	0.17	–
Chlorides (g/L)	143.69	36.85	16.65	13.41	10.26
Sulfates (g/L)	27.91	11.89	5.1	4.23	5.83
Carbonates (g/L)	0.126	0.149	0.227	0.293	–
Ammonium (mg/L)	–	–	<0.17	–	0.2
Nitrates (mg/L)	–	–	<0.20	–	0.15
Phosphates (mg/L)	–	–	–	–	0.05
Fluoride (mg/L)	–	–	0.4	0.4	1.15
Iron (mg/L)	–	< 1.0	–	–	0.05
Silica (mg/L)	–	–	4.4	–	–
Lithium (mg/L)	–	9.8	4.8	0.4	–

4.4 Biological Limnology

This section consists of two subsections dealing separately with the Mar Chiquita Lake lowstand and highstand periods, taking into consideration that (a) the lake's salinity level differed substantially from hypersaline to mesosaline conditions (see Sect. 4.3 and Chap. 3) and (b) that these contrasting situations have a substantial influence in the diversity and dynamics of the lake's biota and the ecosystem ecological dynamics (Hammer 1986).

4.4.1 Lowstand Period

Information on this period is limited, mostly derived from the pioneering, very detailed study on the limnology of Mar Chiquita produced by Seckt (1945). Seck's work was based on microscope observations and included a comprehensive list of the organisms identified together with detailed descriptions and observations. Given the high salinity level at the time of conducted his research (about 250–290 g/L), it is likely that most of the water column was under oxygen-poor conditions even in shallow waters. The list of species identified in Mar Chiquita is given in Appendix 4.1. In the phytoplankton, green algae were dominant. Several species of blue-green algae and diatoms were also abundant. Seckt (1945) found a single species in the zooplankton, the Artemia brine shrimp, a worldwide dispersed genus in continental saline waters. According to the author, "the species occurs in immense quantities as a planktonic species, whereas it appears absent in the mud." The species occurring in Mar Chiquita has been confirmed as *A. franciscana* (Papeschi et al. 2000).

With regard to benthos, Seckt (1945) states that blue-green algae and diatoms were the most abundant groups, whereas green algae were much less abundant in benthos than in plankton. In the organic black mud sediment, Seckt remarks the dominance of heterotrophic, sulfur-related bacteria, "which indicates anoxic conditions, high availability of organic matter and therefore sulfur and nitrogen compounds." The author also remarks that the Mar *Chiquita* benthos was relatively diverse regarding bacteria and algae but with a complete absence of animal species. Seckt concludes his comments by stating that "the Mar Chiquita biocenosis is by no way poor in terms of the abundance of organisms, but it is so in terms of species diversity."

Although not mentioned by Seckt, it is very likely that two more species were present in Mar Chiquita during the lowstand period: the brine fly *Ephydra* sp. (Diptera) and the boatman side bug *Trichocorixa mendozana*. Species of both genera are adapted to high levels of salinity and are known to occur in large numbers in highly saline environments, for example, the Great Salt Lake (Collins 1980; Wurtsbaugh 1992). Supporting evidence for the occurrence of both species in Mar Chiquita during the lowstands comes from observations made by Kanter (1932), who reports "fly larvae that are able to exist with no difficulty in salt water" (very likely *Ephydra sp.*) and also "an aquatic bug (*Notonecta* sp.)" (probably a misidentified *Trichocorixa*). In addition, both species were reported later during the early highstand period (Bucher and Herrera 1981).

4.4.2 Highstand Period

The relative lower salinity conditions generated during this period were reflected primarily on the species diversity present in the lake, including the presence of an invasive fish species, apparently for the first time in the whole geological history of the lake. This section includes a review of the main changes recorded during this period.

4.4.2.1 Prokaryotes

No detailed information is available on the presence of prokaryota organisms for this period, except a survey of functional metabolic groups detected in the lake. The study was based on culturing techniques using selective media for each metabolic group (Bucher and Abril 2006; Abril et al. 2010). Sampling included water and bottom sediment, taken in both pelagic and coastal sites. At the time of sampling (November 2002), the water level was 71.20 m a.s.l. and salinity was 26 g/L. In water, the dominant groups included heterotrophic (both aerobic and anaerobic) and nitrogen-fixing organisms. Their abundance changed along the water-depth gradient, indicating anoxic conditions near the bottom, particularly at over 4 m in depth.

Nitrogen-fixing organisms were abundant in all samples, including both autotrophic and heterotrophic. Nitrogen fixation has been reported in many saline lakes, being probably associated with abundance of cyanobacteria of the genus *Nodularia*, also present in Mar Chiquita (Jonkers et al. 2003). Ammonifiers were dominant in all samples in the pelagic area, peaking at the deepest sites, and also reached high values in organic mud sediments close to the shore. Ammonia oxidizers were not detected in the main lake. Instead, they were recorded on coastal sites, unexpectedly in sediments close to the water surface. According to Jonkers et al. (2003), the presence of ammonia oxidizers is possible in illuminated mats under specific conditions. Nitrate reducers reached the highest values in the open lake water and were scarce in the organic mud sediments. Sulfur-oxidizing organisms were abundant in open waters and in sandy coastal sediments. Contrarily, sulfur reducers were found exclusively in sediments of the more profound sites.

In summary, the functional metabolic group analysis provided new evidence on the importance of sulfur-based metabolic processes in the Mar Chiquita ecosystem, resulting from decreased oxygen solubility in salt water and the high sulfur content of the water.

4.4.2.2 Eukaryotes

Information available for the highstand period was obtained by a detailed survey conducted in Mar Chiquita in July 2005 and January 2006 (Bucher and Abril 2006) (Appendix 4.1). Dense littoral communities of filamentous algae developed, which were not present during the lowstand period – listed in Seckt (1945). These algal masses were dominated by the chlorophyte *Cladophora fracta* (Reati et al. 1997). The widgeon grass (*Ruppia maritima*, Ruppiaceae) invaded the Mar Chiquita Lake during this highstand period, becoming widespread and common (Reati et al. 1997). No vascular aquatic plants had been recorded in the lake during the lowstand period.

The zooplankton species diversity increased during the highstands as compared to the lowstands. However, diversity remained relatively low if compared with freshwater lakes (Wetzel 2001). Dominant groups included Foraminifera, Rotifera, and Crustacea (Appendix 4.1).

Foraminifera *Ammonia beccarii* was very common in plankton and sediments. The occurrence of mostly marine taxa such as Foraminifera in continental lakes is well documented worldwide. In the case of Mar Chiquita, a possible source of dispersal would be the large numbers of several migratory bird species visiting Mar Chiquita, of which some make stops in seashores along their migration routes (Chap. 7).

Rotifera *Brachionus plicatilis* was very abundant. *Brachionus angularis* and *Hexarthra fennica* were also present in low numbers.

Copepoda Three species were common: *Apocyclops procerus*, *Boeckella poopoensis*, and *Cletocamptus deitersi*.

Branchiopoda The population of the Artemia brine shrimp (*Artemia franciscana*) in Mar Chiquita dropped rapidly in Mar Chiquita in response to the increase in water level at the end of the 1970s. In 1998, just very few specimens and scarce accumulations of cysts were found in the lake (Papeschi et al. 2000). Two factors appear critical in determining this population decline: first, a marked lower salinity, well below the Artemia optimal range of 120–200 g/L (Wear et al. 1986; Abatzopoulus et al. 2002), and, second, the intense predation by the silverside fish (*Odontesthes bonariensis*) that invaded the lake during the highstands (see the following section).

Neston With regard to multicellular swimming organisms, *Ephydra* sp. and *Trichocorixa mendozana* continued present in the lake, although at moderate levels.

4.4.2.3   Silverside Fish Invasion

A particularly significant event during the highstand period was the invasion of the lake by the silverside fish (*Odontesthes bonariensis*). This species was introduced in the Mar Chiquita tributary rivers since the 1940s and from there expanded into the lake in the early 1980s, at the time when the lake water salinity fell below 50 g/L. The silverside remained in the lake until 2009 when salinity went over 50 g/L and the lake population became extinct (Bucher and Etchegoin 2006). Later in 2016–2017 the species reinvaded the lake during a new period of high rainfall and decreased salinity (see details in Chap. 5).

4.4.2.4   Benthos Biodiversity

Information on the benthos biodiversity for the highstand period is found in Bucher and Abril (2006). The list of species identified during the highstand period is included in Appendix 4.1. Dominant species included *Ammonia beccarii* (Foraminifera), very abundant, and *Cyprideis* sp. (Ostracoda), found in all samples but particularly abundant in the black mud sediments. Among copepods, *Boeckella poopoensis* (Calanoida) was rare; *Cletocamptus deitersi* (Harpacticoida) was frequent, but not abundant. The snail *Heleobia* sp. (Gasteropoda) was seasonally abundant, particularly common in coastal algae masses. In summary, the benthic fauna in the highstand survey was dominated by foraminifers and ostracods, a frequent pattern in saline lakes (Hammer 1986). Species diversity was higher with respect to Seckt (1945) observations during the lowstands, although still much reduced if compared with freshwater environments (Wetzel 2001).

4.5 Comments

4.5.1 Trophic Condition

The trophic condition of the Mar Chiquita lake was assessed by Pilati et al. (2016) using the Trophic State Index (TSI) (Carlson 1977), based on phosphorous and water transparency. The authors found TSI values higher than 70 in all samples, indicating a clear hypereutrophic condition. They also found that the TSI based on chlorophyll was in the 50–70 range, consistent with a eutrophic condition bordering hypertrophy. The eutrophic condition found in Mar Chiquita is frequent in saline lakes (Hammer 1986).

4.5.2 Black Mud Generation in the Lowstand Periods

In Mar Chiquita, the black mud was abundant during the lowstand period before the late 1970s, when water salinity was over 200 g/L. The mud banks were widespread at low depth even near shores, attracting tourists interested in the therapeutic properties of the mud, as documented by newspapers and magazines of those years. Later, when water salinity fell below 40 g/L at the beginnings of the 1980s, the black mud banks disappeared rapidly from the coastal area, being found only in deep depocenters far from the coast (Durigneaux 1978; MCGlue et al. 2015). This drastic change led the health tourism business to close.

The documented abundance of the black mud during the lowstand period somehow contradicts conclusions of detailed paleolimnological studies in Mar Chiquita, based on the multiproxy and stable isotopic analysis of sediment cores (Piovano et al. 2002, 2004). In summary, the authors propose a conceptual model that identifies evaporite mineral-rich and organic matter-poor sediments as indicators of lowstand periods and organic matter-rich muds with low content of carbonate minerals as signals of highstand periods. This hypothesis is based on the concept that lowstand periods are characterized by low primary productivity. However, the observed presence of large, superficial deposits of organic mud during the lowstands requires further analysis. One important factor to keep in mind is that the saline waters of the lowstands were highly productive, as indicated by the enormous Artemia biomass reported by Seckt (1945) (see Sect. 4.3). In fact, saline lakes are characterized for being low in biodiversity but highly productive (Hammer 1986; Wurtsbaugh et al. 2017).

In Mar Chiquita, the lack of black mud evidence in the core samples corresponding to the lowstand periods when the mud was in fact abundant could be due to the oxidation and remineralization of the organic mud in the sediments between deposition and later collection. This situation could have emerged during re-exposition of the sediments to open air during exceptional and temporary low water level periods. This possibility is especially worth considering given that variations in the water level of Mar Chiquita Lake, which fluctuated within a 10 m range between 1940 and 1990 (Chap. 3). Alternatively, and considering that oxygen solubility decreases exponentially with increases in salt content (Wetzel 2001), the fall in water salinity during the highstands may have allowed oxidizing conditions reaching the lake bottom at a much greater depth than during the lowstand periods.

4.5.3   The Role of Top Predators in Controlling Phytoplankton Density

During a lowstand period, Pilati et al. (2016) reported the simultaneous occurrence in late spring of a marked fall in chlorophyll and peak values in zooplankton, dominated almost entirely by the Artemia brine shrimp *Artemia franciscana*. They also found a year-long negative correlation between Artemia density and chlorophyll biomass, which suggests a strong grazing pressure by the Artemia brine shrimp on phytoplankton. This assumption is also supported by the fact that brine shrimp may ingest about 50–94% of the algae biomass available in the Great Salt Lake water column (Wurtsbaugh et al. 2017).

According to Scheffer (2004), this scenario of clear-water phase that often occurs in freshwater lake plankton at the end of the spring is typical of eutrophic lakes and also appears more likely in lakes with few of no planktivorous fish, allowing a substantial growth of planktivorous invertebrates. Both conditions applied to Mar Chiquita during the study reported by Pilati et al. (2016). This species interaction is further complicated during the highstand period when the silverside fish (*Odontesthes bonariensis*) invaded the lake (Bucher and Etchegoin 2006) and Artemia became almost extinct. The details regarding the specific response of phytoplankton to this change in top predators remains an open question until new fluctuations of the lake level provide another opportunity for further research.

Appendix 4.1

List of plankton and benthos species identified in the Mar Chiquita Lake, Córdoba, Argentina. Sources: (1): Seckt (1945); (2): Reati et al. (1997); (3): Cohen (1998); (4): Bucher and Abril (2006)

Year	1945	1989	1998	2005–2006
Salinity (g/L)	100	35	45	35
Source	(1)	(2)	(3)	(4)
Cyanophyta (cyanobacteria)				
Anabaena spiroides		X		X
Anabaenopsis circularis			X	X
Calothrix parietina				
Chrococus turgidus	X			
Clastidium setigerum	X			
Cyanobium spp.				X
Dermacarpa prasina	X			
Gomphosphaeria sp.				X
Lyngbya aestuarii	X			
Lyngbya hieronymusii	X			
Lyngbya sp.	X		X	X
Merismopedia sp.				X
Microcoleus vaginatus	X			
Microcystis chroococcoidea	x			
Microcystis sp.				
Nodularia spumigera		X		X
Nostoc commune	X			
Nostoc gelatinosum	X			
Oscillatoria (Limnothrix) putrida	X			
Oscillatoria limosa	X			
Oscillatoria sp.		X	X	X
Phormidium sp.				
Phormidium breve	X			
Phormidium deflexum	X			
Phormidium murrayi	X			
Phormidium simplicissimum	X			
Phormidium chalybeum	X			
Pseudanabaena constricta	X			
Siphonema polonicum	X			
Tolypothrix sp.				X
Trichodesmium iwanoffianum	X			
Heterokontophyta (diatomeas)				
Achnanthes delicatula	X			X
Amphora aff. exigua			X	

(continued)

Appendix 4.1 (continued)

Year	1945	1989	1998	2005–2006
Salinity (g/L)	100	35	45	35
Source	(1)	(2)	(3)	(4)
Amphora copulata			X	
Amphora normanii	X			
Amphora veneta				
Amphora sp.				X
Anomoneis sphaerophora	X			X
Aulacoseira sp.				X
Campylodiscus bicostatus			X	
Campylodiscus clypeus		X		
Campylodiscus sp.				X
Chaetoceros muelleri		X		
Chaetocerus sp.			X	X
Chanantes brevipes			X	
Cocconeis pediculus	X			
Cocconeis placentula			X	
Cocconeis sp.				X
Craticula sp.				X
Cyclotella choctawhatcheeana			X	
Cyclotella meneghiniana	X			
Cyclotella spp.				X
Cymatopleura sp.				X
Cymbella pusilla			X	
Cymbella ventricosa	X			
Cymbella spp.				X
Diatoma sp.				X
Entomoneis alata		X		
Entomoneis sp.			X	X
Fragilaria sp.				X
Gyrosigma acuminatum	X			
Gyrosigma macrum	X			
Gyrosigma strigilis			X	
Gyrosigma sp.				X
Hantzschia amphioxys	X			
Hyalodiscus sp.				X
Melosira lineata		X		
Melosira moniliformis			X	
Melosira sp.		X		X
Navicula cincta			X	
Navicula halophila	X			
Navicula mutica			X	
Navicula placentula	X			
Navicula pygmaea	X			
Navicula recens			X	

Year	1945	1989	1998	2005–2006
Salinity (g/L)	100	35	45	35
Source	(1)	(2)	(3)	(4)
Navicula salinarum	X			
Navicula sp.		X		X
Nitzschia acicularis	X			
Nitzschia capitellata			X	
Nitzschia hustedtiana			X	
Nitzschia linearis	X			
Nitzschia scalpelliformis			X	
Nitzschia spp.				X
Pinnularia sp.				X
Pleurosira sp.				X
Rhopalodia spp.				X
Surirella striatula			X	
Surirella sp.				X
Synedra tabulata	X			
Synedra ulna		X		
Tabularia fasciculata			X	
Thalassiosira sp.				X
Ulnaria sp.				X
Xanthophyta (algas verde-amarillas)				
Vaucheria sp.		X		
Chlorophyta (algas verdes)				
Binuclearia tetraona		X		
Bulbochaete sp.	X			
Chlamydomonas sp.				X
Cladophora fracta		X		
Closterium pronum		X		X
Enteromorpha salina	X			
Enteromorpha intestinalis	X			
Enteromorpha intestinalis		X		
Mougeotia sp.	X			
Oocystis spp.				X
Pediastrum spp.				X
Rhizoclonium hieroglyphicum	X			
Scenedesmus spp.				X
Stigeoclonium falklandicum	X			X
Ulothrix limnetica	X			
Ulothrix pseudoflacca		X		
Ulothrix tenuis	X			
Ulotrichopsis sp.				X
Euglenophyta (euglenas)				
Euglena spp.				X
Trachelomonas sp.				X

(continued)

Appendix 4.1 (continued)

Year	1945	1989	1998	2005–2006
Salinity (g/L)	100	35	45	35
Source	(1)	(2)	(3)	(4)
Dinophyta (dinoflagelados)				
Peridinium spp.			X	
Sarcomastigophora (protozoarios)				
Foraminiferida (foraminiferos)				
Ammonia beccarii parkinsoniana		X		
Ammonia sp.				X
Spermatophyta (*plantas superiores*)				
Ruppiaceae				
Ruppia maritima		X		X
Rotifera (rotíferos)				
Brachionus sp.			X	X
Brachionus plicatilis				X
Brachionus angularis				X
Brachionus fennica				X
Crustacea (crustáceos)				
Branchiopoda				
Artemia franciscana	X	X	X	
Ostracoda		X		
Cyprideis? sp.				X
Copepoda				
Calanoida		X	X	
Boeckella poopoensis				X
Cyclopoida				
Apocyclops procerus				
Harpacticoidea				
Cletocamptus deitersi				X
Insecta (insectos)				
Diptera				
Ephydridae				
Ephydra? sp.	X			
Dolichopodidae				
Hydrophorus praecox		X		
Especie indeterminada				X
Hemiptera				X
Corixidae				
Molusca (caracoles)				
Hidrobiidae				
Heleobia (*Littoridina*) sp.		X		X
Ceratodes sp.				

References

Abatzopoulus T, Beardmore J, Clegg J, Sorgeloos P (2002) Artemia. Basic and applied biology. Kluwer Academic Publishers, London

Abril A, Noe L, Merlo C (2010) Grupos metabólicos microbianos de la laguna Mar Chiquita (Córdoba, Argentina) y su implicancia en el ciclado de nutrientes. Ecol Aust 20(1):81–88

Bucher EH (ed) (2006) Bañados del Rio Dulce y Laguna Mar Chiquita (Córdoba. Argentina. Academia Nacional de Ciencias (Córdoba, Argentina)

Bucher EH, Abril AB (2006) Limnología Biológica. In: Bucher EH (ed) Bañados del Rio Dulce y Laguna Mar Chiquita (Córdoba, Argentina). Academia Nacional de Ciencias, Córdoba, pp 117–137

Bucher EH, Bucher AE (2006) Limnología Física y Química. In: Bucher EH (ed) Bañados del Rio Dulce y Laguna Mar Chiquita (Córdoba, Argentina). Academia Nacional de Ciencias (Córdoba, Argentina), Córdoba, pp 79–191

Bucher EH, Etchegoin M (2006) El pejerrey como recurso. In: Bucher EH (ed) Bañados del Río Dulce y Laguna Mar Chiquita (Córdoba, Argentina). Academia Nacional de Ciencias (Córdoba, Argentina), Córdoba, pp 201–217

Bucher E, Herrera G (1981) Comunidades de aves acuáticas de la Laguna Mar Chiquita (Córdoba, Argentina). Ecosur 8:91–120

Bucher EH, Stein AF (2016) Large salt dust storms follow a 30-year rainfall cycle in the Mar Chiquita Lake (Córdoba, Argentina). PLoS One 11(6):e0156672

Carlson R (1977) A trophic state index for lakes. Limnol Oceanogr 22:361–369

Cohen RG (1998) Prospección y evaluación del recurso natural Artemia(Crustacea, Branchiopoda) en formade quistes en la Republica Argentina y su utilidad en proyectos de acuicultura para Latinoamérica. Unpublished Technical Report. Facultad de Ciencias Exactas, Fisicas y Naturales, Universidad de Buenos Aires.Buenos Aires, Argentina

Collins N (1980) Population ecology of Ephydra cinerea Jones (Diptera, Ephydridae), the only benthic metazoan of the Great Salt Lake, U.S.A. Hydrobiologia 68(2):99–1122

Durigneaux J (1978) Composición química de las aguas y barros de la Laguna Mar Chiquita en la provincia de Córdoba. Academia Nacional de Ciencias (Córdoba, Argentina), Córdoba

Hammer UT (1986) Saline Lake ecosystems of the world. Dr. W. Junk Publishers, Boston

Jonkers H, Ludwig R, De Wit R, Pringault O, Muy-zer G, Niemann H, Finke N, De Beer D (2003) Structural and functional analysis of a microbial mat ecosystem from a unique permanent hypersaline in-land Lake: La Salada de Chiprana (NE Spain). FEMS Microbiol Ecol 44:175–189

Kanter H (1932) La cuenca cerrada de la Mar Chiquita en el norte de la Argentina. Bol Acad Nac Cienc Córdoba (Argentina) 22:285–232

Martinez DE (1991) Caracterización geoquímica de las aguas de la Laguna Mar Chiquita, provincia de Córdoba Dr Thesis, Universidad Nacional de Córdoba

McGlue MM, Ellis GS, Cohen AS (2015) Modern muds of Laguna Mar Chiquita (Argentina): particle size and organic matter geochemical trends from a large saline lake in the thick-skinned Andean foreland. In: Larsen D, Egenhoff SO, Fishman NS (eds) Paying attention to mudrocks: priceless! The Geological Society of America (Boulder, Colorado), Special Paper 15, p 1–18

Oren A (2002) Halophilic microorganisms and their environments. Kluver Academic Publishers, London

Papeschi AG, Cohen RG, Pastorino XI, Amat F (2000) Cytogenetic proof that the brine shrimp Artemia franciscana (Crustacea, Branchiopoda) is found in Argentina. Hereditas 133:159–166

Pilati A, Marcellino M, Bucher E (2016) Nutrient, chlorophyll and zooplankton seasonal variations on the southern coast of a subtropical saline lake (Mar Chiquita, Córdoba, Argentina). Int J Limnol 52:263–271

Piovano EL, Ariztegui D, Damatto Moreiras S (2002) Recent environmental changes in Laguna Mar Chiquita (central Argentina): a sedimentary model for a highly variable saline lake. (Sedimentology, 49.ina): a little known, secularly fluctuating saline lake. Int J Salt Lake Res 5:187–219

Piovano EL, Ariztegui D, Bernasconi SM, McKenzie JA (2004) Stable isotopic record of hydrological changes in subtropical Laguna Mar Chiquita (Argentina) over the last 230 years. Holocene 14:525–535

Reati G, Florin M, Fernandez G, Montes C (1997) The Laguna de Mar Chiquita (Córdoba, Argentina): a little known, secularly fluctuating saline Lake. Int J Salt Lake Res 5:187–219

Scheffer M (2004) Ecology of Shallow Lakes. Kluwer Academic Publishers, Dordrecht, The Netherlands

Seckt H (1945) Estudios hidrobiológicos hechos en la Mar Chiquita. Bol Acad Nac Cienc (Córdoba, Argentina) 37:279–309

Thorpe S (2004) Langmuir circulation. Annu Rev Fluid Mech 36:55–79

Wallace Gwynn J (2002) Great Salt Lake: an overview of change. Utah Geological Survey, Salt Lake City

Wallace Gwynn J (2004) What causes the foam on Great Salt Lake? http://geology.utah.gov/survey-notes/gladasked/gladfoam.htm. Retrieved 12 Nov 2005

Wear R, Haslett S, Alexander N (1986) Effects of temperature and salinity on the biology of Artemia franciscana Kellogg from Lake Grassmere, New Zealand. 2. Maturation, fecundity, and generation times. J Exp Mar Biol Ecol 98(1–2):167–183

Wetzel RG (2001) Limnology. Lake and Rivers ecosystems. Academic, New York

Wurtsbaugh N (1992) Food-web modification by an invertebrate predator in the Great Salt Lake. Oecologia 89(2):168–175

Wurtsbaugh WA, Miller C, Null SE, DeRose Justin R, Wilcock P, Hahnenberger M, Howe F, Moore J (2017) Decline of the world's saline lakes. Nat Geosci 10(11):816–821

Chapter 5
Fish

5.1 Introduction

The fish fauna from Mar Chiquita includes three clearly differentiated regions in terms of structural and hydrological characteristics and species diversity: the lake, the Rio Dulce wetlands, and the Primero and Segundo rivers. In total, the fish fauna of Mar Chiquita wetland contains 37 species. The Rio Dulce wetland has 31 species, whereas Primero and Segundo rivers are home to 17 and 15 species, respectively. The Dulce River shares 18 species with Primero River and Segundo River, whereas Primero and Segundo rivers share 10 species. Only nine species are common to the three rivers (Haro 2006). See Table 5.1 for a complete list of the fish of Mar Chiquita.

There is only one species endemic to the region, the mojarra fish *Astyanax cordovae*, which is restricted to the Primero and Segundo rivers (Haro 2006). There are also three exotic species: pejerrey silverside (*Odontesthes bonariensis*), common carp (*Cyprinus carpio*), and Gambusia (*Gambusia affinis*) (Haro 2006).

As a whole, the fish fauna of the Mar Chiquita wetland is clearly related to the Paranaense fish fauna of the Paraguay, Parana, and Rio de la Plata basin, whereas it lacks species of Andean origin (Menni 2004). The Paranaense character of the Mar Chiquita ichthyofauna can be related to the old connections that once existed between the tributaries to Mar Chiquita and the Parana River basin (Kröhling and Iriondo 1999; Brunetto and Iriondo 2007), and also to the relatively recent temporary connection between the Salado and the Dulce rivers (Bucher et al. 2006) (see also Chaps. 1 and 11). On the contrary, the basins of Primero and Segundo rivers, with origin in the Cordoba Mountains, would have been isolated since the Holocene (Iriondo and Garcia 1993; Kröhling and Iriondo 1999), which might explain the presence of one species endemic to the region.

© Springer Nature Switzerland AG 2019
E. H. Bucher, *The Mar Chiquita Salt Lake (Córdoba, Argentina)*,
https://doi.org/10.1007/978-3-030-15812-5_5

Table 5.1 List of fish species occurring in Mar Chiquita according to habitat and biotopes. From Haro (2006)

Habitat/species	Local name	Dulce river	Primero and Segundo rivers	Mar Chiquitalake
Open waters				
Salminus maxillosus	dorado	Frequent		
Prochilodus lineatus	sábalo	Abundant		
Odontesthes bonariensis[a]	pejerrey	Rare	Rare	Rare-abundant[b]
River bottom				
Cyprinus carpio[a]	carpa	Frequent	Rare	
Otocinclus vittatus	viejita enana	Rare		
Loricariichthys melanocheilus	vieja del agua	Rare		
Hypostomus commersoni	vieja del agua	Frequent		
Hypostomus cordovae	vieja del agua		Frequent	
Trachelyopterus striatulus	torito	Frequent		
Pimelodus albicans	bagre blanco	Frequent	Frequent	
Pimelodella gracilis	desconocido	Rare		
Pimelodella laticeps	bagre cantor	Frequent	Frequent	
Rhamdia quelen	bagre sapo	Rare	Rare	
Heptapterus mustelinus	bagre anguila		Rare	
Still waters				
Parodon tortuosus	boguita		Rare	
Cyphocharax voga	sabalito		Rare	
Leporinus obtusidens	boga	Frequent		
Characidium sp.	desconocido	Rare		
Aphyocharax erythrurus	mojarrín	Rare		
Astyanax bimaculatus	mojarra	Frequent	Frequent	
Astyanax cordovae	mojarrón		Rare	
Astyanax eigenmanniorum	mojarra cola roja	Frequent	Frequent	
Astyanax fasciatus	mojarra	Rare		
Bryconamericus iheringi	fina	Rare	Frequent	
Oligosarcus jenynsii	dientudo	Rare		
Cheirodon interruptus	mojarrita	Rare	Frequent	
Odontostilbe microcephala	mojarrín	Frequent	Rare	
Serrasalmus spilopleura	piraña	Frequent		
Hoplias malabaricus	tararira	Frequent	Frequent	
Jenynsia multidentata	orillero	Frequent	Rare	Frequent
Gambusia affinis[a]	orillero	Rare	Rare	Rare
Cnesterodon decemmaculatus	orillero	Rare	Frequent	Rare
Cichlasoma facetum	palometa	Rare	Rare	
Eigenmannia virescens	banderita	Rare		
Temporary ponds				
Synbranchus marmoratus	anguila	Rare	Rare	
Corydoras paleatus	amarillito	Rare	Frequent	
Hoplosternum littorale	cascarudo	Rare		

[a]Introduced species
[b]The population size of the silverside pejerrey depends on the lake water salinity

5.2   Mar Chiquita Lake

The number of species present in the lake is very low due to the limiting factor represented by the high salinity of its water (Chap. 4). Before the 1970s, because of the high water salinity, the fish fauna of the lake was restricted to only three species of small fish belonging to the superfamily Poeciliidae (Cyprinodontiformes), known as "livebearers" (viviparous) or "nearshores." In Mar Chiquita they are restricted to the proximity of the lakeshore and absent from open waters.

Rio de la Plata one-sided livebearer (*Jenynsia multidentata*, Anablepidae): A species widely dispersed in South America, commonly seen in shallow ponds in large numbers. It can tolerate high salinity conditions. In Mar Chiquita, it has been observed to tolerate a salinity level of 50 g/L.

Livebearer (*Cnesterodon decemmaculatus*, Poeciliidae): Uncommon in Mar Chiquita, resistant to salinity, although to a lesser degree than *Jenynsia multidentata*. This species is widely used in the laboratory for toxicity assessment studies.

Western mosquitofish (*Gambusia affinis*, Poeciliidae): A nonnative species, introduced in Mar Chiquita for mosquito control, with similar characteristics of the two previously mentioned species. These three species show a marked convergence in several characters, including small size (less than 10 cm), sexual dimorphism (females considerably larger than males), and internal fertilization and viviparity (livebearing). Also, the three species are euryhaline in habitat, being common in salt marshes and other saline habitats (Haro 2006). During the highstand period that started in the 1970s, the pejerrey silverside invaded the lake, becoming extremely abundant (see Sect. 5.5).

5.3   Dulce River Wetland

This subregion is the richest regarding fish biodiversity, with 31 species being found in the terminal reaches of the river and the numberless associated ponds and marshes in the Dulce River floodplain. The Dulce River wetland is also characterized by the presence of a large number of medium-sized fish, with species that reach or exceed 1 kg in weight at the adult stage. It is also interesting to note the presence of the *glass knifefish* (*Eigenmannia virescens*) (Sternopygidae), locally known as *banderita*, a weakly electric freshwater fish related to species of the Amazon River basin (Haro 2006).

5.3.1   Habitat Diversity and Species Communities

The fish species occurring in the Mar Chiquita area may be grouped in the following habitat categories (based on (Menni 2004): mainstream open waters, river bottom, still waters, and highly saline environments.

Mainstream Open Waters Dominated by large-sized species (relative to the rest), including *dorado* (*Salminus maxillosus*, Characidae), a powerful predator; *sábalo* (*Prochilodus lineatus*, Prochilodontidae), an illiophagous species that sucks and eats organic mud, for which its mouth is especially adapted; and *boga* (*Leporinus obtusidens*, Anostomidae), an omnivorous species that feeds on fruits and seeds, water plants, crabs, and river snails.

River Bottom Includes species in the catfish group of the Siluriformes order. In Mar Chiquita there are species with armor-plated skins of the Loricariidae family (genera *Hypostomus* and *Otocinclus*) locally known as "*viejas del agua*," which feed mostly on mud and algae, and naked species, mostly zoophagous, locally known as "*bagres*." These species belong to the families Auchenipteridae (genera *Trachipterus* and *Rhamdia*) and Pimelodidae (genera *Pimelodus*, *Pimelodella*, and *Heptapterus*).

Still Waters Fish community composed mainly of species of less than 20 cm in length, *silvery*, of compressed body and forked caudal fin, and which feed mostly on invertebrates and fish eggs. The commonest species in Mar Chiquita are the mojar-ras *Astyanax bimaculatus* and *A. eigenmanniorum*. Also, two larger-sized ambush fish predators occur in this habitat: the *tararira* wolf fish (*Hoplias malabaricus*), which may reach up to 55 cm, and the well-known piranha (*Serrasalmus specularis*) of about 25 cm.

5.3.2   Factors Influencing Species Diversity

The greatest diversity of species and functional types that occur in the Dulce River is associated with the following factors: (a) great water flow; (b) great habitat diversity due to the presence of a large floodplain with multiple ponds and marshes, generated and maintained by the annual flooding of the Dulce River (Bistoni et al. 1992); and (c) the fact that the Dulce River basin is located in more tropical areas than those of the Primero and Segundo rivers (Chap. 3).

5.4 Primero and Segundo Rivers

The fish fauna recorded in both Primero and Segundo rivers up to 10 km upstream from their mouths in Mar Chiquita Lake amounts to 17 and 15 species, respectively. Both are low-flow rivers and lack the variety of biotopes characteristic of the Dulce River (Bistoni et al. 1992; Haro 1996).

5.5 The Invasive Pejerrey Silverside

The substantial increase of the water level followed by a similarly significant decrease in salinity (Chaps. 3 and 4) provided a suitable habitat for the pejerrey silverside (*Odontesthes bonariensis*, Atherinopsidae), which invaded the lake through the tributary rivers and expanded rapidly all over the extended lake (Bucher and Etchegoin 2006).

The pejerrey silverside is a Neotropical freshwater species highly tolerant to brackish environments, native to the inland waters of Buenos Aires province in Argentina (Mancini and Grossman 1995). It belongs to the atherinid group, which includes several marine species and only two species that occur in terrestrial environments, including the pejerrey silverside. The latter two species have origin in marine species that colonized the terrestrial environments of eastern Argentina, as a consequence of sea level variations during the Pleistocene. The marine background of the pejerrey may explain its wide habitat range, which includes freshwater, seawater, and saline marshes and lakes (Tejedor 2002). In Mar Chiquita, it has been observed at salt concentrations up to 50 g/L, the highest value recorded for the species (Menni 2004). The pejerrey has been introduced for sport and commercial fishing purposes in other Argentine provinces and several countries in South America and other continents (Mancini and Grossman 1995).

The pejerrey silverside is basically zoophagous, with food resources varying with age and size of the specimens. Individuals of less than 200 mm feed on zooplankton, including crustaceans, insects, snails and clams, and small fish. Individuals of more than 200 mm have a predominantly piscivorous diet, with a lower proportion of clams and insects. The piscivorous diet is mainly cannibalistic, since pejerrey silverside is the only fish present in the open lake waters (Bucher and Etchegoin 2006).

The existence of extended cannibalism, as occurs in Mar Chiquita, may affect the stability of populations in the lake. According to Claessen and De Roos (2003), cannibalism may induce regular cycles or irregular variations in the population without tending to a single equilibrium point. Further research is necessary to elucidate this issue.

5.5.1 Population Dynamics and Salt Tolerance

The pejerrey silverside fish was first detected in Mar Chiquita in 1979 after an unprecedented algae bloom in the late 1970s when water salinity dropped below 45 g/L. The fish population grew rapidly and reached very high levels.

After 2003, a water level decrease period started in the lake, and the pejerrey population declined sharply in 2008 when water salinity reached 50 g/L and was no longer detected when salinity exceeded 56 g/L (Bucher and Etchegoin 2006).

Between 2017 and 2018, the silverside invaded the lake again during a short period when the lake level increased, and the water salinity dropped below 42 g/L and became almost extinct about 8 months later when the lake level receded, and salinity rose over 52 g/L.

In summary, the long-term records obtained in Mar Chiquita indicate that the upper limit of salinity tolerance for a sustainable population of the silverside ranges between 45 and 50 g/L. Above this level, and up to about 56 g/L, adult individuals may survive, but reproduction appears unviable (Bucher and Etchegoin 2006). This limit is much higher than what was previously estimated according to field observations in other water bodies in Argentina (Gomez et al. 2007) and experimental data (Tsuzuki et al. 2000).

5.5.2 The Silverside as a Resource

Following the substantial increase in the silverside population that started in the late 1970s, the province of Córdoba authorized commercial fishing in 1981. The new resource was very productive during the initial years (1981–1986). In the following years, there was a considerable increase in the number of active fishermen and fishing pressure, leading to overproduction and a marked fall in the market price. In 1988 the local authorities attempted to correct this problem and organize the pejerrey silverside industry under a more regulated system (Bucher and Etchegoin 2006). Unfortunately, there was no systematic data recording of this activity, including productivity and fishing, even less population assessment.

At this respect, the only reliable source of information available consists of the detailed records obtained by two fishermen during the 1992–1999 period; those data were processed and published by Bucher and Etchegoin (2006). In summary, these reports provided the following valuable data (Fig. 5.1):

1. The average annual catch per fisherman ranged between 1 and 8 mT. In 6 of the 8 years, it reached between 7 and 8 mT, in 2 was below 4 tons, and in a single year, less than 2 tons.
2. In 5 of the 8 years analyzed, the annual catch of the two fishermen taken together was between 13 and 16 tons/year; in 2, between 6 and 8; and in 1, between 2 and 3.

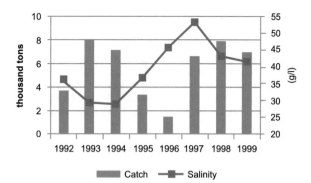

Fig. 5.1 Annual total silverside fish catch by two fishermen in the Mar Chiquita Lake during the 1992–1999 period, related to water salinity

Overall, the pejerrey silverside invasion showed the species potential as a significant economic resource for Mar Chiquita, at least during the highstand periods when the water level was above 69 m a.s.l. However, implementing an efficient, large-scale exploitation system appears difficult, given the unpredictability of the climate, the size of the lake and the required boat fleet, and transportation costs to the distant markets. A simpler, low-scale opportunistic exploitation scheme implemented whenever the pejerrey silverside reaches a suitable population level seems a much more feasible approach.

References

Bistoni M, Haro J, Gutierrez M (1992) Ictiofauna del Río Dulce en la provincia de Córdoba (Argentina). Iheringia 72:105–111

Brunetto E, Iriondo MH (2007) Neotectónica en la Pampa Norte (Argentina). Rev Soc Geol Esp 20:17–29

Bucher EH, Etchegoin M (2006) El pejerrey como recurso. In: Bucher EH (ed) Bañados del río Dulce y laguna Mar Chiquita (Córdoba, Argentina). Academia Nacional de Ciencias (Córdoba, Argentina), Córdoba, pp 201–217

Bucher E, Marcellino A, Ferreyra C, Molli A (2006) Historia del Poblamiento Humano. In: Bucher E (ed) Bañados del Rio Dulce y Laguna Mar Chiquita (Córdoba, Argentina). Academia Nacional de Ciencias (Córdoba, Argentina), Córdoba, pp 301–325

Claessen D, De Roos A (2003) Bistability in a size-structured population model of cannibalistic fish – a continuation study. Theor Popul Biol 64:49–65

Gomez S, Menni R, Gonzalez Naya J, Ramirez L (2007) The physical–chemical habitat of the Buenos Aires pejerrey, *Odontesthes bonariensis* (Teleostei, Atherinopsidae), with a proposal of a water quality index. Environ Biol Fish 78:161–171

Haro JG (2006) Peces. In: Bucher EH (ed) Bañados del Rio Dulce y Laguna Mar Chiquita. Academia Nacional de Ciencias de Córdoba, Córdoba

Iriondo M, Garcia N (1993) Climatic variations in the Argentine plains during the last 18,000 yr. Palaeogeogr Paleoclimatol Palaeoecol 101:209–220

Kröhling D, Iriondo M (1999) Upper Quaternary Palaeoclimates of the Mar Chiquita Area, North Pampa, Argentina. Quat Int 57(58):149–163

Mancini M, Grossman M (1995) El Pejerrey de las lagunas pampeanas. Universidad Nacional de
    Rio Cuarto and Universidad Nacional del Centro de la Provincia de Buenos Aires, Rio Cuarto
Menni R (2004) Peces y ambientes en la Argentina continental. Museo Argentino de Ciencias
    Naturales de Buenos Aires, Buenos Aires
Tejedor D (2002) El pejerrey como recurso genético. In: Grossman M (ed) Fundamentos biológi-
    cos, económicos y sociales para una correcta gestión del recurso pejerrey. Astyana, Azul
Tsuzuki MY, Aikawa H, Strussmann CA, Takashima F (2000) Comparative survival and growth of
    embryos, larvae and juveniles of pejerrey (*Odontesthes bonariensis* and *O. hatcheri*) at different
    salinities. J Appl Ichthyol 16:126–130

Chapter 6
Amphibians and Reptiles

6.1 Introduction

This chapter summarizes the available information on the herpetofauna of the Mar Chiquita wetland's region, including updated data from published literature and field observations (E.H. Bucher). Given the great environmental heterogeneity characteristic of Mar Chiquita, amphibian and reptile distribution is not homogeneous. At least two broad types of environments can be distinguished for these groups in terms of habitat suitability: (a) Chaco forest of the peripheral area around the Mar Chiquita depression and (b) the Dulce River wetland, including the transition woodland, halophytic scrubland, flooded savanna plant formations, and permanent and temporary water courses of the Dulce River (Chap. 10).

The complete list of species in the Mar Chiquita wetland and the transition to the Chaco forest woodland is given in Tables 6.1, 6.2, and 6.3. A detailed description of those species that occur in the Dulce River wetland is included in the text. No endemisms occur in the Mar Chiquita wetland in any of the groups described in this chapter.

6.2 Amphibians

The total number of amphibian species present in the Mar Chiquita Lake and the Dulce River wetland amounts to 16 and is distributed in three families (Leptodactylidae with 11 species, Bufonidae with 3 species, and Hylidae with 2 species) (Table 6.1). There are no amphibian species endemic to Mar Chiquita (Cei 1980; Leynaud et al. 2006). The species occurring in the wetland area are listed below (see also Table 6.1).

© Springer Nature Switzerland AG 2019
E. H. Bucher, *The Mar Chiquita Salt Lake (Córdoba, Argentina)*,
https://doi.org/10.1007/978-3-030-15812-5_6

Table 6.1 List of amphibians occurring in the Mar Chiquita wetland and in the surrounding Chaco forest area. From Leynaud et al. (2006)

Order/family	Species	Chaco forest	Dulce River wetland
Anura			
Leptodactylidae			
	Ceratophrys cranwelli	x	
	Lepidobatrachus asper	x	x
	Leptodactylus bufonius	x	
	Leptodactylus chaquensis	x	x
	Leptodactylus gracilis	x	
	Leptodactylus laticeps	x	x
	Leptodactylus latinasus	x	x
	Leptodactylus mystacinus	x	x
	Odontophrynus americanus		x
	Physalaemus biligonigerus	x	x
	Pleurodema tucumana	x	
Bufonidae			
	Rhinella arenarum	x	x
	Rhinella fernandezae		x
	Rhinella schneideri	x	x
Hylidae			
	Phyllomedusa sauvagii	x	
	Scinax nasicus	x	

Lepidobatrachus asper: Inhabits dry scrubland and semiarid areas, including saline environments. Breeds in temporary pools and even in water tanks on cattle farms. During the dry season, these frogs burrow themselves underground, only to emerge again after the rainy season.

Leptodactylus chaquensis: Found in Chaco woodlands, near ponds and flooded areas. Hides in burrows or in dense vegetation. Eggs are laid in large foam nests over puddles and flooded areas, in water less than 15 cm deep.

Leptodactylus laticeps: Found in semiarid woodland. Commonly hides in rodent burrows. Breeds in temporary ponds. Eggs are laid in foam masses inside burrows. The skin secretion is toxic to humans. Listed as a threatened species by the IUCN.

Leptodactylus latinasus: Its natural habitat is subtropical and tropical grasslands near standing waterbodies. Hides and breeds in underground chambers or under fallen trees. Eggs are laid in foam masses.

Leptodactylus mystacinus: Occurs mostly in grasslands. Some populations occur in forests and deforested areas. Its preferred habitat is standing waterbodies in grasslands. Eggs are laid in foam masses in burrows underground.

Odontophrynus americanus: Inhabits open grasslands and savannas. It is a fossorial species that remains buried most of the year. During the breeding season, it emerges and is found in shallow, temporary ponds and flooded areas. Eggs are laid in the muddy bottom, and the tadpoles develop rapidly in the same site before the ponds dry.

Physalaemus biligonigerus (ranita llorona): Adapted to a wide variety of habitats, including floodplains, margins of rivers, streams, and ponds, and even in urban environments. Builds foam nests. Calls while floating on the water. Eggs are laid in spherical foam nests that float on the water surface.

Rhinella arenarum: Prefers small ponds or bogs with stagnant water, in dry, temperate environments, mostly in open areas. Also common in urban areas. Oviposition occurs in temporary water bodies. Eggs are laid inside long, jelly strings that remain attached to the submerged vegetation. In winter, it finds refuge in burrows or burrows itself underground.

Rhinella fernandezae: Common in flooded grasslands. Breeds in temporary ponds. Lays eggs inside long, jelly strings attached to vegetation.

Rhinella schneideri (Rococó, sapo buey): A large-sized toad (about 22 cm). Preferred habitats include woodland savannas and saline wetlands. Breeds in permanent and temporary ponds, preferring those with sparse vegetation. Lays eggs inside long, jelly strings attached to vegetation.

6.3   Reptiles

The total number of reptile species occurring in Mar Chiquita amounts to 36 (18 snakes, 13 lizards, 3 amphisbenas, and 2 turtles) (Tables 6.2 and 6.3). As with amphibians, there are no species endemic to the area (Cabrera 1996, 2004; Leynaud and Bucher 1999; Leynaud et al. 2006).

Table 6.2 List of snake species occurring in the Mar Chiquita wetland and in the surrounding Chaco forest area. From Leynaud et al. (2006)

Order/family	Species	Chaco forest	Rio Dulce wetland
Serpentes			
Leptotyphlopidae			
	Leptotyphlops melanotermus		
Boidae			
	Boa constrictor occidentalis		x
	Epicrates cenchria alvarezi	x	
Colubridae			
	Boiruna maculata	x	x
	Liophis anomalus	x	x
	Liophis poecilogyrus	x	x
	Lystrophis dorbignyi		x
	Lystrophis pulcher	x	x
	Oxyrhopus rhombifer bachmanni	x	
	Phalotris bilineatus	x	
	Philodryas patagoniensis	x	x
	Philodryas psammophidea	x	x
	Sibynomorphus turgidus		x
	Waglerophis merremii	x	x

(continued)

Table 6.2   (continued)

Order/family	Species	Chaco forest	Rio Dulce wetland
Elapidae			
	Micrurus pyrrhocryptus	x	x
Viperidae			
	Bothrops alternatus		x
	Bothrops neuwiedi diporus	x	
	Crotalus durissus terrificus	x	

Table 6.3 List of lizard, turtle, and amphisbaenian species recorded in Mar Chiquita. From Leynaud et al. (2006)

Order/family	Species	Chaco forest	Wetland
Squamata (lizards)			
Anguidae			
	Ophiodes intermedius	x	x
Gekkonidae			
	Homonota fasciata	x	
Polychrotidae		x	
	Leiosaurus paronae	x	
	Urostrophus gallardoi	x	
Tropiduridae		x	
	Liolaemus chacoensis	x	
	Stenocercus doellojuradoi	x	
	Tropidurus etheridgei	x	
	Tropidurus spinulosus	x	
Scincidae		x	
	Mabuya dorsivittata	x	x
Teiidae		x	x
	Teius oculatus	x	x
	Teius teyou	x	x
	Tupinambis merianae	x	x
	Tupinambis rufescens	x	
Amphisbaenia		x	
(worm lizards)			
Amphisbaenidae		x	
	Amphisbaena bolivica	x	
	Amphisbaena darwinii	x	x
	Anops kingii	x	
Testudinea (turtles)		x	
Testudinidae		x	
	Chelonoidis chilensis	x	
Chelidae (side-neck turtles)		x	
	Phrynops hilarii	x	

6.3.1 Snakes

The snakes present in Mar Chiquita belong to five families: Leptotyphlopidae (1), Boidae (2), Colubridae (11), Elapidae (1), and Viperidae (3).

Leptotyphlops melanotermus: A blind, fossorial, and small snake (0.40 m) in length. Widely distributed in Argentina.

Boa constrictor (Lampalagua). A large terreestrial boa (up to 3.5 m). Common in Chaco woodland.

Epicrartes cenchria (Boa arcoiris) (1.5 m). Terrestrial boa (1.5m). In Chaco woodland.

Boiruna maculata (Musurana): A large, black snake (1.80 m), widely distributed in Argentina. Preys mostly on snakes, also on frogs, toads, lizards, birds, and mammals.

Liophis anomalus: Grass snake (0.70 m). Found near rivers, lakes, and swamps. Feeds on insects, toads, and frogs.

Liophis poecilogyrus: Grass snake (0.70 cm). Occurs in grasslands and wetland areas (streams, lakes, and wetlands), also in forests and even in urban areas. Feeds on toads, frogs, and fish.

Lystrophis dorbignyi: A fer-de-lance mimic (nonvenomous) (0.80 m). Prefers grasslands, margins of water bodies, and open forests. Feeds on frogs and toads.

Lystrophis pulcher: Coral snake mimic (nonvenomous) (0.70 m). Common in dry woodlands. Burrows itself in soft or sandy soils. Feeds on frogs, toads, and lizards.

Oxyropus rhombifer: Falsa coral (1.00 m) A coral snake mimic. Feeds mostly on snakes.

Phalotrix bilineatus: Falsa coral. A coral snake mimic. (0.40 m). Prefers grasslands.

Philodryas patagoniensis: Large-sized (1.50 m) grass snake. Widely distributed throughout Argentina, in open areas (grasslands, open forests, peri-urban areas, etc.). Feeds on various food resources: lizards, birds, mammals, and even snakes. One of the most frequent species in the Mar Chiquita wetlands.

Philodryas psammophideus: Grass snake (1 m). Widely distributed in Argentina, mainly in grasslands. Feeds principally on lizards. One of the commonest species in the Mar Chiquita wetlands.

Sibynomorphus turgidus: Small-sized (40 cm) fer-de-lance mimic (nonvenomous). Occurs in open, humid areas. Feeds on insect larvae, snails, and other small prey.

Waglerophis merremii: A fer-de-lance mimic (nonvenomous) (0.90 m). Widely distributed in Argentina, prefers humid environments and grasslands near rivers, streams, and flooded areas. Feeds on amphibians (especially frogs and toads).

Micrurus pyrrhocryptus: A coral snake (poisonous) (1.30 m). Occurs in forests, woodlands, and grasslands. Feeds on snakes, small lizards, and amphisbaenas.

Bothrops alternatus: Yarará grande (1.80 m). A fer-de-lance snake. Occurs mostly in wetland grasslands. One of the commonest species in the Rio Dulce Wetland. Poisonous.

Bothrops neuwiedi: Yarará chica (1.10 m). Common, ubiquitous. Poisonous.

Crotalus durissus: Cascabel (1.80 m). A rattlesnake that prefers dry Chaco woodland. Poisonous.

Fig. 6.1 The fer-de-lance snake (known locally as "*yarará*") (*Bothrops alternatus*). Poisonous. One of the most prevalent species in the Rio Dulce wetlands

6.3.1.1   Snakes of Medical Importance

In the Mar Chiquita and Dulce River wetlands, there are four venomous snake species of sanitary importance: two fer-de-lance species, *Bothrops alternatus* (yarará grande or víbora de la cruz) and *Bothrops neuwiedi* (yarará chica); one rattlesnake, *Crotalus durissus* (cascabel); and one coral snake, *Micrurus pyrrhocryptus* (víbora de coral). *Bothrops alternatus* (Fig. 6.1) is the commonest and the cause of most bite accidents.

6.3.2   Lizards

The fauna of lizards recorded in the area comprises 13 species distributed in six families: Anguidae (1), Gekkonidae (1), Polychrotidae (2), Tropiduridae (4), Scincidae (1), and Teiidae (4) (Table 6.2). Of all the species present in the region, only *Mabuya dorsivittata*, *Teius teyou*, and *Tupinambis merianae* occur in the Dulce River wetland, whereas the remaining species are restricted to the transitional peripheral area (Table 6.3).

This last group includes the following species:

Ophiodes intermedius: A semi-fossorial, legless lizard (0.27 m). With only vestigial back legs of less than 10 mm, it is frequently mistaken for snakes. Common in wetlands.

Teius oculatus: A bright green, medium-sized lizard (0.12 m). In central Argentina, it occurs in humid biotopes including wetlands.

Tupinambis merianae (Argentine black and white tegu): A large lizard (up to 1.30 m). Present in central, northern, and eastern Argentina in open areas, forests, grasslands, and wetlands. It is a very good swimmer, capable of remaining submerged for more than 20 min. An omnivorous species, feeds on a very diverse diet, including small animals, fish, eggs, insects, snails, carrion, and fruits.

6.3.3 Turtles

Only two turtle species are present in the Mar Chiquita wetland (Cabrera 1996).

Phrynops hilarii (Hilaire's toadhead turtle) is an aquatic species and occurs in low numbers, in the tributary rivers close to the Mar Chiquita Lake, but not in the more saline waters of the main lake.

Chelonoidis chilensis (Chaco tortoise) is a terrestrial species that occurs in semi-arid woodland vegetation in the periphery of the Masr Chiquita depression, but not in the wetland.

6.3.4 Amphisbaenians

These are limbless, subteranean lizards. Three rare and poorly known species occur in the Mar Chiquita wetland (Table 6.3).

References

Cabrera M (1996) In: di Tada IE, Bucher EH (eds) Biodiversidad de la Provincia de Córdoba. Universidad Nacional de Rio Cuarto, Rio Cuarto

Cabrera MR (2004) Las serpientes de Argentina Central. Universidad Nacional de Córdoba, Córdoba

Cei JM, 2:609 (1980) Amphibians of Argentina. Monitore Zoologico Italiano 2:1–609

Leynaud GC, Bucher EH (1999) La fauna de serpientes del Chaco sudamericano: diversidad, distribución geográfica y estado de conservación. Academia Nacional de Ciencias (Córdoba, Argentina), Córdoba

Leynaud GC, Pelegrin N, Lescano JN (2006) In: Bucher EH (ed) Bañados del Rio Dulce y Laguna Mar Chiquita. Academia Nacional de Ciencias (Córdoba, Argentina), Córdoba, pp 219–235

Chapter 7
Birds

7.1   Introduction

The Mar Chiquita wetland, which includes the Mar Chiquita Lake and the Dulce River wetlands, is characterized by a rich and diverse bird fauna. The lake supports 148 species of aquatic birds, some of which reach very high numbers given the exceptional dimensions of the lake. Besides its value for the conservation of bird biodiversity in general terms, the lake is particularly important for three species of flamingos and intercontinental migratory shorebirds. The present analysis is focused on the bird species primarily associated with the aquatic environments of Mar Chiquita Lake and Dulce River wetlands. Species present in the highlands surrounding the wetland are not included. The complete list of species occurring in the region can be found in Torres and Michelutti (2001).

7.2   Species Diversity and Abundance

The total of bird species directly associated with Mar Chiquita wetland amounts to 148 (Table 7.1). No endemic species have been recorded in this group. Of those species, 37 (26%) are long-range migrants from North America, Northern South America, High Andes, and Patagonia.

North America: 28 species that nest in North America during the boreal summer reach Mar Chiquita during the austral summer (September–March). The shorebirds (20 species) are the most abundant intercontinental migrants (see details below). Other common North American migrants include one heron, little blue heron (*Egretta caerulea*); one duck, blue-winged teal (*Anas discors*); one bird of prey (osprey, *Pandion haliaetus*); one gull (Franklin's gull, *Larus pipixcan*); two terns, black tern (*Chlidonias niger*) and South American tern (*Sterna hirundo*);

© Springer Nature Switzerland AG 2019   73
E. H. Bucher, *The Mar Chiquita Salt Lake (Córdoba, Argentina)*,
https://doi.org/10.1007/978-3-030-15812-5_7

Table 7.1 List of aquatic birds recorded in the Mar Chiquita wetland. Data from Torres and Michelutti (2006) and Toledo et al. (2018)

Family/scientific name	Common English name	Local common name
Rheidae		
Rhea Americana	*Greater rhea*	*Ñandú*
Podicipedidae		
Podilymbus podiceps	*Pied-billed grebe*	*Macá pico grueso*
Tachybaptus dominicus	*Least grebe*	*Maca chico*
Rollandia Rolland	*White-tufted grebe*	*Macá común*
Podiceps major	*Great grebe*	*Macá grande*
Podiceps occipitalis	*Southern Silvery grebe*	*Macá plateado*
Anhingidae		
Anhinga anhinga	*Anhinga*	*Bigua vibora*
Phalacrocoracidae		
Phalacrocorax brasilianus	*Neotropical Cormorant*	*Biguá*
Anhimidae		
Chauna torquata	*Southern screamer*	*Chajá*
Ardeidae		
Ardea cocoa	*Cocoi heron*	*Garza mora*
Tigrisoma lineatum	*Rufescent tiger heron*	*Hoco colorado*
Botaurus pinnatus	*Pinnated bittern*	*Mirasol grande*
Ixobrychus involucris	*Stripped-backed bittern*	*Mirasol común*
Ardea alba	*Great white egret*	*Garza blanca*
Egretta thula	*Snowy egret*	*Garcita blanca*
Bubulcus ibis	*Cattle egret*	*Garcita bueyera*
Egretta caerulea	*Little blue heron*	*Garza azul*
Syrigma sibilatrix	*Whistling heron*	*Chiflón*
Butorides striatus	*Green-backed heron*	*Garcita azulada*
Nycticorax nycticorax	*Black-crowned night heron*	*Garza bruja*
Threskiornitidae		
Phimosus infuscatus	*Bare-faced ibis*	*Cuervillo cara pelada*
Plegadis chihi	*White-faced ibis*	*Cuervillo de cañada*
Harpiprion caerulescens	*Plumbeous ibis*	*Bandurria mora*
Theristicus caudatus	*Buff-necked ibis*	*Bandurria boreal*
Theristicus melanopis	*Black-faced ibis*	*Bandurria austral*
Platalea ajaja	*Roseate spoonbill*	*Espatula rosada*
Ciconiidae		
Mycteria Americana	*Wood stork*	*Tuyuyú*
Ciconia maguari	*Maguari stork*	*Cigueña americana*
Jabiru mycteria	*Jabiru*	*Yabiru*
Phoenicopteridae		
Phoenicopterus chilensis	*Chilean flamingo*	*Flamenco austral*
Phoenicoparrus andinus	*Andean flamingo*	*Parina grande*
Phoenicoparrus jamesi	*Puna flamingo*	*Parina chica*

(continued)

Table 7.1   (continued)

Family/scientific name	Common English name	Local common name
Anhimidae		
Chauna torquata	Southern screamer	Chajá
Anatidae		
Cairina moschata	Muscovy duck	Pato criollo
Sarkidiornis melanotus	Comb duck	c
Dendrocygna bicolor	Fulvous whistling duck	Siriri colorado
Dendrocygna autumnalis	Black-bellied whistling duck	Siriri vientre negro
Amazonetta brasiliensis	Brazilian duck	Pato colorado
Dendrocygna viduata	White-faced whistling duck	Siriri pampa
Coscoroba coscoroba	Coscoroba swan	Cisne coscoroba
Cygnus melancoryphus	Black-necked swan	Cisne cuello negro
Chloephaga melanoptera	Andean goose	Guayata
Anas sibilatrix	Southern wigeon	Pato overo
Anas discors	Blue-winged teal	Pato medialuna
Anas bahamensis	White-cheeked pintail	Pato gargantilla
Anas georgica	Yellow-billed pintail	Pato maicero
Anas flavirostris	Speckled teal	Pato barcino
Anas platalea	Red shoveler	Pato cuchara
Anas cyanoptera	Cinnamon teal	Pato colorado
Anas versicolor	Silver teal	Pato capuchino
Callonetta leucophrys	Ringed teal	Pato de collar
Netta peposaca	Rosy-billed pochard	Pato picazo
Netta erythrophthalma	Southern pochard	Pato castaño
Heteronetta atricapilla	Black-head duck	Pato cabeza negra
Oxyura vittata	Lake duck	Pato zambullidor chico
Nomonyx dominicus	Masked duck	Pato fierro
Accipitridae		
Elanus leucurus	White-tailed kite	Milano blanco
Accipiter striatus	Sharp-shinned hawk	Esparvero común
Rostrhamus sociabilis	Snail kite	Caracolero
Rupornis magnirostris	Roadside hawk	Taguató común
Buteogallus meridionalis	Savanna hawk	Aguilucho colorado
Geranoaetus albicaudatus	White-tailed hawk	Aguilucho alas largas
Rallidae		
Laterallus melanophaius	Rufous-sided crake	Burrito común
Coturnicops notate	Speckled crake	Burrito enano
Porzana flaviventer	Yellow-breasted crake	Burrito amarillo
Pardirallus sanguinolentus	Plumbeous rail	Gallineta común
Pardirallus maculatus	Spotted rail	Gallineta overa
Gallinula melanops	Spot-flanked gallinule	Polla pintada
Gallinula chloropus	Common gallinule	Polla de agua
Fulica leucoptera	White-winged coot	Gallareta chica
Fulica armillata	Red-gartered coot	Gallareta ligas rojas

(continued)

Table 7.1 (continued)

Family/scientific name	Common English name	Local common name
Fulica rufifrons	*Red-fronted coot*	*Gallareta escudete rojo*
Porphyrula martinica	*Purple gallinule*	*Polla sultana*
Aramides ypecaha	*Giant wood rail*	*Ipacaá*
Aramidae		
Aramus guarauna	*Limpkin*	*Carau*
Jacanidae		
Jacana jacana	*Wattled jacana*	*Jacana*
Rostratulidae		
Nycticryphes semicollaris	*South American painted-snipe*	*Aguatero*
Recurvirostridae		
Himantopus melanurus	*South American stilt*	*Tero real*
Charadriidae		
Charadrius collaris	*Collared plover*	*Chorlito de collar*
Charadrius falklandicus	*Two-banded plover*	*Chorlito doble collar*
Pluvialis squatarola	*Black-bellied plover*	*Chorlo artico*
Pluvialis dominica	*American golden plover*	*Chorlo dorado*
Vanellus chilensis	*Southern lapwing*	*Tero común*
Charadrius semipalmatus	*Semipalmated plover*	*Chorlito palmado*
Charadrius modestus	*Rufous-chestered dotterel*	*Chorlo pecho rojizo*
Oreopholus ruficollis	*Tawny-throated dotterel*	*Chorlo cabezon*
Scolopacidae		
Limosa haemastica	*Hudsonian godwit*	*Becasa de mar*
Calidris alba	*Sanderling*	*Playerito blanco*
Calidris bairdii	*Baird's sandpiper*	*Playerito unicolor*
Calidris ferruginea	*Curlew sandpiper*	
Calidris fuscicollis	*White-rumped sandpiper*	*Playrito rabadilla blanca*
Calidris melanotos	*Pectoral sandpiper*	*Playerito pectoral*
Calidris canutus	*Red knot*	*Playero rojizo*
Calidris pusilla	*Buff-breasted sandpiper*	*Playerito enano*
Phalaropus tricolor	*Wilson's phalarope*	*Falaropo común*
Tringa solitaria	*Solitary sandpiper*	*Pitotoy solitario*
Tringa flavipes	*Lesser yellowlegs*	*Pitotoy chico*
Tringa melanoleuca	*Greater yellowlegs*	*Pitotoy Grande*
Bartramia longicauda	*Upland sandpiper*	*Batitu*
Actitis macularia	*Spotted sandpiper*	*Playerito manchado*
Catoptrophorus semipalmatus	Willet	*Playero ala blanca*
Micropalam himantopus	*Stilt sandpiper*	*Playero zancudo*
Tringites subruficollis	*Buff-breasted sandpiper*	*Playerito canela*
Arenaria interpres	*Ruddy turnstone*	*Vuelve piedra*
Gallinago gallinago	*Common snipe*	*Becasina comun*
Limnodromuys griseus	*Short-billed dowitcher*	*Becasa gris*

(continued)

Table 7.1 (continued)

Family/scientific name	Common English name	Local common name
Laridae		
Larus maculipennis	Brown-hooded gull	Gaviota capucho café
Larus cirrocephalus	Grey-headed gull	Gaviota capucho gris
Larus scoresbi	Dolphin gull	Gaviota gris
Larus dominicanus	Kelp gull	Gaviota cocinera
Larus pipixcan	Franklin's gull	Gaviota chica
Larus atlanticus	Olorg's gull	Gaviota cangrejera
Chlidonias niger	Black tern	Gaviotin negro
Stercorarius parasiticus	Parasitic jaeger	Salteador chico
Phaetusa simplex	Large-billed tern	Ati
Sterna nilotica	Gull-billed tern	Gaviotin pico grueso
Sterna hirundo	Common tern	Gaviotin golondrina
Sterna trudeaui	Snowy-crowned tern	Gaviotín lagunero
Sterna superciliaris	Yellow-billed tern	Gaviotín chico común
Rynchops niger	Black skimmer	Rayador
Caprimulgidae		
Eleothreptus anomalus	Sickle-winged nightjar	Atajacaminos de pantano
Alcedinidae		
Megaceryle torquata	Ringed kingfisher	Martín pescador grande
Chloroceryle amazon	Amazon kingfisher	Martin pescador mediano
Chloroceryle americana	Green kingfisher	Martin pescador chico
Furnaridae		
Cinclodes fuscus	Buff-winged cinclodes	Remolinera comun
Cinclodes comechingonus	Comechingones Cinclodes	Remolinera serrana
Phleocryptes melanops	Wren-like rushbird	Junquero
Spartonoica maluroides	Bay-capped wren-spinetail	Espartillero enano
Certhiaxis cinnamomea	Yellow-throated spinetail	Curutie colorado
Tyrannidae		
Pseudocolapteryx sclateri	Crested doradito	Doradito copetón
Pseudocolapteryx dinelianus	Dinelli's doradito	Doradito pardo
Pseudocolopteryx flaviventris	Warbling doradito	Doradito común
Fluvicola pica	Pied water tyrant	Viudita blanca
Fluvicola albiventer	Black-backed water tyrant	Viudita blanca
Lessonia rufa	Rufous-backed negrito	Sobrepuesto común
Hymenops perspicillatus	Spectacled tyrant	Pico de plata
Polystictus pectoralis	Bearded tachuri	Tachuri canela
Tachuris rubrigastra	Many-colored rush tyrant	Sietecolores de laguna
Hirundinidae		
Riparia riparia	Bank swallow	Golondrina zapadora
Hirundo rustica	Barn swallow	Golondrina tijerita
Troglodytidae		
Cistothorus platensis	Grass wren	Ratona aperdizada

(continued)

Table 7.1 (continued)

Family/scientific name	Common English name	Local common name
Emberizidae		
Sporophila collaris	*Rusty-collared seedeater*	*Dominó*
Agelaius cyanopus	*Unicolored blackbird*	*Varillero negro*
Agelaius ruficapillus	*Chestnut-capped blackbird*	*Varillero congo*
Agelastius thilius	*Yellow-winged blackbird*	*Varillero ala amarilla*
Pseudoleistes virescens	*Brown-and-yellow marshbird*	*Pecho amarillo común*
Amblyramphus holosericeus	*Scarlet-headed blackbird*	*Federal*

and two swallows, bank swallow (*Riparia riparia*) and barn swallow (*Hirundo rustica*).

Northern South America: The migratory race of the black skimmer (*Rynchops niger*) visits Mar Chiquita in summer, joining the local, permanent race *R. n. intercedens* (Torres and Michelutti 2001).

Andean Puna Highlands: The altitudinal migrations between the High Andes region known as Puna Andina (which encompasses part of Argentina, Bolivia, Chile, and Peru), include the three species of flamingos present in Mar Chiquita (Sect. 7.3).

Patagonia: Includes birds that nest in the south of South America and spend the winter season (March to October) in Mar Chiquita. Six species occur in Mar Chiquita: three seagulls, kelp gull (*Larus dominicanus*), Olrog's gull (*Larus atlanticus*), and dolphin gull (*Larus scoresbii*), one ibis (buff-necked Ibis, *Theristicus melanopis*), and two shorebirds (see details below).

7.3   Flamingos

Of the six extant species of flamingos in the world, three occur in Mar Chiquita: the Chilean flamingo (*Phoenicopterus chilensis*), the Andean flamingo (*Phoenicoparrus andinus*), and the Puna flamingo (*Phoenicoparrus jamesi*).

7.3.1   The Chilean Flamingo

The most abundant flamingo species in Mar Chiquita, seen all year round, with peak numbers during the breeding season and much lower numbers during winter (Bucher et al. 2000; Bucher and Curto 2012). The breeding cycle usually starts in November, when birds concentrate in large numbers in traditional sites, usual islands with wide mudflat beaches (Figs. 7.1 and 7.2). Nestlings abandon the islands and start to congregate in crèches in early March, reaching peak numbers from April to mid-May. In April, both adults and juveniles start to leave the colony (Bucher 2006) (Figs. 7.1 and 7.2).

Fig. 7.1 Aerial view of a breeding colony of Chilean flamingo in Mar Chiquita. Mar Chiquita is the main breeding area of the whole species' distribution range. (From Bucher 2006)

Fig. 7.2 A large crèche of young Chilean flamingos in Mar Chiquita. (From Bucher 2006)

Mar Chiquita is a very important breeding site for the Chilean flamingo. No doubt, its breeding colonies are the most spectacular bird concentrations observed in Mar Chiquita. A long-term study on the breeding of the Chilean flamingo throughout 42 years of continuous observations provided interesting information on the Chilean flamingo breeding behavior (Bucher and Curto 2012).

During the study period (1969–2010), the lake underwent great oscillations of water salinity, which ranged between 297 and 25 g L^{-1} according to the water level and lake area (Bucher and Bucher 2006). As a result, the ecological conditions of the lake changed substantially, particularly in terms of water salinity, distribution of islands and mudflats, and presence and abundance of Artemia brine shrimp (*Artemia franciscana*) during lowstands and pejerrey silverside fish (*Odontesthes bonariensis*) during the highstands (Chaps. 3, 4, and 5).

Flamingos bred irregularly during both high- and low-salinity periods (11 successful attempts in 42 years). The observed irregularity in breeding confirmed a well-known tendency for flamingos to breed sporadically in large colonies, usually in a few places to which they tend to be faithful (Johnson and Cézilly 2007). The observed frequency of successful breeding attempts in Mar Chiquita (once every 3.85 years) is very similar to that recorded for the closely related greater flamingo (*Phoenicopterus roseus*) during a 40-year period in Etosha National Park, Namibia (once every 3.64 years), and during a 26-year period at Fuente de Piedra, Spain (once every 4.33 years) (Johnson and Cézilly 2007).

Concerning the possible influence of the studied factors on flamingo breeding in Mar Chiquita, comparison of breeding and non-breeding years showed that availability of mudflat islands devoid of mammal predators is the only habitat characteristic invariably associated with successful breeding events. It was also found that breeding success may be affected by sudden variations in water level, which may flood the nests during the incubation and early nesting periods. Rapid decreases in lake level due to droughts may also affect breeding by establishing connections between the islands and the mainland, allowing predation by terrestrial vertebrates, particularly foxes. Other factors (water level, water salinity, local rainfall, and presence of Artemia and the pejerrey silverside fish) were within similar ranges in breeding and non-breeding years.

Besides environmental factors, it is likely that food availability may represent an additional critical factor to be considered. Unfortunately, food availability is difficult to measure in Mar Chiquita, given the size of the lake and the possibility that flamingos may feed in other wetlands, particularly in the nearby Dulce River wetlands.

Of particular interest is the fact that, according to the long-term breeding records obtained in Mar Chiquita, flamingos can survive and breed even at times when the Artemia brine shrimp (*Artemia franciscana*), considered the flamingos' preferred food, is not present in the lake (Bucher and Curto 2012). This finding confirms research results in the Camargue (France) showing that flamingos are typical opportunistic breeders, capable of exploiting highly unpredictable and short-lived resources that appear over the nesting region (Bechet and Johnson 2008; Johnson and Cézilly 2007). Moreover, our observations in Mar Chiquita also suggest that the black organic mud present in saline wetlands may provide an important staple food to flamingos (particularly flight-less juveniles in crèches close to the breeding colony), as proposed by Jenkin (1957).

Regarding population numbers, the average number of birds in the colonies in Mar Chiquita was about 6000 juveniles and at least about 100,000 adults. The record number was 42,800 juveniles and about 100,000 adults (Bucher et al. 2000; Bucher and Curto 2012).

7.3.2 The Andean Flamingo and the Puna Flamingo

Both species occur in lower numbers, mostly as winter visitors. They breed in salt lakes at a high elevation of about 3000–4000 m in the Andes of Argentina, Bolivia, Chile, and Peru. At least part of their populations migrates to lower wetlands for the winter, reaching Mar Chiquita as winter visitors. According to Torres and Michelutti (2006), the numbers recorded in the lake are in the order of 6000 Andean and 2000 Puna flamingos.

7.4 Shorebirds

Mar Chiquita is one of the most important sites in South America in terms of inland shorebird diversity and abundance, which has justified its nomination as a site of Hemispheric Importance by the Western Hemisphere Shorebird Reserves Network (WHSRN). To date, 30 species have been recorded (Charadriidae, 9; Scolopacidae, 18; Jacanidae, 1; Rostratulidae, 1; and Recurvirostridae, 1 (Table 7.1)). Although the populations may vary markedly between years, the most common species in the region include southern lapwing (*Vanellus chilensis*) and South American stilt (*Himantopus melanurus*) among residents and greater yellowlegs (*Tringa melanoleuca*), lesser yellowlegs (*Tringa flavipes*), white-rumped sandpiper (*Calidris fusci-collis*), and Wilson's phalarope (*Phalaropus tricolor*) among North America migrants. Wintering species from Patagonia include the rufous-chested dotterel (*Charadrius modestus*) and the tawny-throated dotterel (*Oreopholus ruficollis*). Both occur in very low numbers. Recently, a group of vagrant individuals of the Eurasian species curlew sandpiper (*Calidris ferruginea*) were recorded in Mar Chiquita, for the first time in Argentina (Toledo et al. 2018).

With regard to the Wilson's phalarope, Mar Chiquita is one of the main wintering areas of this species in South America, where record numbers of between 250,000 and 500,000 individuals have been reported in some years (Nores 2011).

These numbers represent a significant proportion of the global population, estimated at between 1 and 1.5 million birds (Lesterhuis and Clay 2009). However, the abundance of Wilson's phalarope has gone through great oscillations among years, particularly during the highstand period, probably due to variations in food availability (Lesterhuis and Clay 2009).

7.5   Swans

Mar Chiquita is a home to the two South American swan species: the black-necked swan (*Cygnus melancorryphus*) and the coscoroba swan (*Coscoroba coscoroba*). Until the mid-1970s, during the lake lowstands, the former species was scarce, whereas the latter was more common, but not abundant (Nores and Izurieta 1980). After the increase of the lake water level, the populations of both species increased gradually, peaking during periods of highest water level (Torres and Michelutti 2006).

7.6   Other Bird Concentrations and Colonies

Besides flamingos and shorebirds, several other species also congregate in large numbers. According to Blanco et al. (2001), in Mar Chiquita white-faced ibis (*Plegadis chihi*), red-gartered coot (*Fulica armillata*), white-winged coot (*Fulica leucoptera*), brown-hooded gull (*Larus maculipennis*), and neotropic cormorant (*Phalacrocorax olivaceus*) exhibit the highest concentrations of the whole southern cone of South America. This statement is confirmed by isolated observations in Mar Chiquita including more than 44,200 coots of both species (July 1992), 25,000 white-faced ibis (1993), and 42,000 neotropic cormorant (July 1994) (Torres and Michelutti 2006).

Concentrations of between 300 and 400,000 snowy egret (*Egretta thula*) were reported by Nores (1986). The silvery grebe (*Podiceps occidentalis*) and the white-tufted grebe (*Rollandia rolland*) may also reach high numbers during winter, being dispersed over the whole lake (E. Curto and E.H. Bucher, personal observations).

Several large multi-species nesting colonies have been observed in the Dulce River wetlands between 1998 and 2001. An outstanding case was a colony that included the following number of breeding pairs: 22,800 neotropic cormorant, 8500 cattle egret (*Bubulcus ibis*), 5750 white-necked heron (*Casmerodius albus*), 3700 black-crowned night heron (*Nycticorax nycticorax*), 2850 snowy egret (*Egretta thula*), and 1350 great egret (*Ardea cocoi*). Another significant record corresponds to a monospecific colony of cattle egret on the southern coast of Mar Chiquita (Segundo River delta), which reached a record number for the species of about 28,000 pairs in December 1996 (Torres and Michelutti 2006).

7.7   The Effect of Lake Water Level Oscillations on Bird Diversity and Abundance

The avifauna of Mar Chiquita includes a combination of freshwater and saltwater wetland species, whose composition and abundance changes markedly with water level and salinity oscillations (Nores 2011). The comparison of observations made along low- and highstands provides an indication of these changes (Nores and Izurieta 1980; Bucher and Herrera 1981; Torres and Michelutti 2001, 2006).

During the lowstand period, the lake is characterized by the abundance of large concentrations of shorebirds, particularly *Phalaropus tricolor* and *Tringa flavipes*, black-winged stilt (*Himantopus melanurus*), flamingo (*Phoenicopterus chilensis*), and gulls (especially *Larus maculipennis*). Besides plovers, other bird species that are usually found only in the sea coasts are attracted by the vast estuaries formed in the river mouths. Piscivorous birds (especially *Phalacrocorax olivaceus* and several heron species) are restricted to the river mouths, where high fish mortality occurs in the contact area between the freshwater and the hypersaline water of the lake. During the highstands, a marked reduction of shorebirds and a moderate decrease of flamingos take place, together with a generalized increase of piscivorous species (cormorants) and herbivorous species (swans, ducks, and coots) (Torres and Michelutti 2006) (Chap. 12).

7.8 Conservation

The Mar Chiquita and Dulce River wetland region is regarded as an important bird area for several threatened birds in the Neotropics (Wege and Long 1995). It is a Ramsar International site and a Hemispheric site of the Western Hemisphere Shorebird Reserve Network (WHSRN); in addition, Mar Chiquita is listed as an AICA (Important Bird Area for conservation) by the Aves Argentinas NGO.

According to the International Union for Conservation of Nature (IUCN), the list of species that is considered in the "vulnerable" category includes the Andean flamingo (*Phoenicoparrus andinus*), *Porzana spiloptera*, and *Larus atlanticus*. Species in the "near threatened" category include the Puna flamingo (*Phoenicoparrus jamesi*), Chilean flamingo (*Phoenicopterus chilensis*), buff-breasted sandpiper (*Tryngites subruficollis*), sickle-winged nightjar (*Eleothreptus anomalous*), bay-capped wren-spinetail (*Spartonoica maluroides*), bearded tachuri (*Polystictus pectoralis*), and Dinelli's doradito (*Pseudocolopteryx dinellianus*).

The eskimo curlew (*Numenius borealis*), a species that occurred in Mar Chiquita in large numbers, is almost certainly extinct. At present, the species is listed by IUCN as "Critically Endangered (Possibly Extinct)." A North American migrant, its last confirmed record in North America was in 1963 and in South America (south of Mar Chiquita) in 1939 (Wetmore 1939). Additional records were reported later, including one in Mar Chiquita in 1991 (Torres and Michelutti 2006), but none were confirmed. Updated and detailed information on the conservation status of birds from Mar Chiquita is included in Torres and Michelutti (2007).

References

Bechet A, Johnson AR (2008) Anthropogenic and environmental determinants of Greater Flamingo *Phoenicopterus roseus* breeding numbers and productivity in the Camargue (Rhone delta, southern France). IBIS 150:69–79

Blanco D, Minotti P, Canevari P (2001) In: Blanco D, Carbonell M (eds) Investigación del valor del Censo Neotropical de Aves Acuáticas como herramienta para la conservación y el manejo de la vida silvestre. El Censo Neotropical de Aves Acuáticas. Los primeros 10 años: 1990–1999. Parte I. Wetlands International, Memphis, pp 1–19

Bucher E (2006) In: Bucher EH (ed) Flamencos. Bañados del río Dulce y laguna Mar Chiquita (Córdoba, Argentina). Academia Nacional de Ciencias de Córdoba, Córdoba, pp 251–261

Bucher E, Bucher AE (2006) In: Bucher E (ed) Limnología Física y Química. Bañados del Rio Dulce y Laguna Mar Chiquita. Academia Nacional de Ciencias, Córdoba, pp 79–101

Bucher EH, Curto E (2012) Influence of long-term climatic changes on breeding of the Chilean flamingo in Mar Chiquita, Córdoba, Argentina. Hydrobiologia 697:127–137

Bucher E, Herrera G (1981) Comunidades de aves acuáticas de la Laguna Mar Chiquita (Córdoba, Argentina). Ecosur 8:91–120

Bucher EH, Echevarria AL, Juri M, Chani JM (2000) Long-term survey of Chilean Flamingo breeding colonies on Mar Chiquita Lake, Córdoba, Argentina. Waterbirds 23:114–118

Jenkin P (1957) The Filter-Feeding and Food of Flamingoes (Phoenicopteri). Philosophical Transactions of the Royal Society, London B 240:410–493

Johnson AR, Cézilly F (2007) The greater flamingo. T & AD Poyser, London

Lesterhuis AJ, Clay R (2009) Conservation plan for Wilson's phalarope (*Phalaropus tricolor*). Manomet Center for Conservation Sciences, Manomet

Nores M (1986) In: Scott DA, Carbonell M (eds) Argentina. Inventario de humedales de la Región Neotropical. WRB and IUCN, Cambridge, UK

Nores M (2011) Long-term waterbird fluctuations in Mar Chiquita Lake, Central Argentina. Waterbirds 34(3):381–388

Nores M, Izurieta D (1980) Aves de ambientes acuáticos de Córdoba y centro de Argentina. Secretaría de Agricultura y Ganadería de Córdoba, Córdoba

Toledo M, Quaglia A, Vergara Tabares D (2018) New sandpiper from an interior sea: confirmation of Curlew Sandpiper (Calidris ferruginea) for Argentina. Rev Bras Ornitologia 26:214–216

Torres R, Michelutti P (2001) Las aves de ambientes acuáticos del sistema Laguna Mar Chiquita – Bañados del Rio Dulce (provincias de Córdoba y Santiago del Estero, Argentina). Bol Acad Nac Cienc (Córdoba, Argentina) 66:61–73

Torres R, Michelutti P (2006) Aves acuáticas. In: Bucher E (ed) Bañados del Rio Dulce y Laguna Mar Chiquita. Academia Nacional de Ciencias (Córdoba, Argentina), Córdoba, pp 237–249

Torres R, Michelutti P (2007) Reserva de Uso Múltiple Bañados del Río Dulce y Laguna Mar Chiquita. In: Di Giacomo AS, De Franceso M, Coconier EG (eds) Áreas importantes para la conservación de las aves en Argentina. Sitios prioritarios para la conservación de la biodiversidad. Asociación Ornitológica del Plata, Buenos Aires, pp 134–137

Wege DC, Long AJ (1995) Key areas for threatened birds in the Neotropics. BirdLife, Cambridge, UK

Wetmore A (1939) Recent observations on the Eskimo Curlew in Argentina. Auk 56:475–476

Chapter 8
Mammals

8.1 Introduction

The mammalian fauna of Mar Chiquita has been poorly studied, and systematic and detailed surveys in the region are therefore lacking. This chapter includes only those species of particular interest that are most significant regarding abundance or ecological characteristics. The selected species are grouped in three main sections: (a) species whose presence in the area has been confirmed; (b) species very likely to occur in the area, but still not confirmed; and (c) species that have become locally extinct after the settlement of European immigrants. Bats (Chiroptera) are not included given the lack of information relative to the Mar Chiquita wetland (Parera 2002; Torres and Tamburini 2018).

8.2 Species of Confirmed Occurrence in Mar Chiquita

Lutrine opossum (*Lutreolina crassicaudata*, Caenolestidae). Local name: *Comadreja colorada.* The species occurs in grassy woodland and marshy areas. Restricted to the Dulce River wetlands in low numbers.

Coypu (*Myocastor coypus*, Myocastoridae) (Fig. 8.1). Local name: *coipo, nutria.* A large, herbivorous, semiaquatic rodent that lives in burrows alongside stretches of water and feeds on river plant stems. Native to subtropical and temperate South America, it has been introduced to North America, Europe, Asia, and Africa, primarily by fur ranchers. Although it is still valued for its fur, at present it is considered a pest in most of its invaded range.

The coypu has been an important natural resource in Mar Chiquita since the Spanish arrival to the region. According to the Jesuit priests that visited the wetlands, the coypu was the staple food of the local natives (Lozano 1754). By the mid-nineteenth century, high demand for coypu fur developed in Europe. Argentina

© Springer Nature Switzerland AG 2019   85
E. H. Bucher, *The Mar Chiquita Salt Lake (Córdoba, Argentina)*,
https://doi.org/10.1007/978-3-030-15812-5_8

Fig. 8.1 Coypu (also known as *nutria*) (*Myocastor coypus*). Very common species, subject to intense hunting and captive breeding until the end of the twentieth century. (From Haro et al. 2006)

was probably one of the first countries to export coypus to Europe. Since then, and for an extended period, the sale of coypu fur and meat became a significant contribution to the economy of the Mar Chiquita region, especially for the locality of Miramar. Exports of wild coypu skins from Argentina peaked during the 1896–1924 period. By the end of 1923, a drastic decrease in the wild population was first noticed, to the point that in 1930 Europe banned the import of wild coypus and required that exporters establish breeding farms to meet the growing demands (Garcia Mata 1973).

This new policy forced the development of coypu captive breeding in Argentina. The Mar Chiquita region, particularly the coastal town of Miramar, played a significant role in the development of captive breeding techniques. According to Griva (1973), it was precisely in Miramar where the present-day domestication of coypu and captive breeding practices were first applied, which later expanded worldwide. In 1938, there were already 200 breeding farms in Miramar, and in 1951, 183,000 skins were produced annually (Curto and Castellino 2006; Griva 1973).

Miramar remained as the only center of coypu captive breeding at the beginning of the twenty-first century. Early in the century, production of coypu meat for human consumption was incorporated; this meat has the added value of being lean and low in cholesterol. Still in 2005, between 8000 and 11,000 coypus were produced annually (Curto and Castellino 2006). However, in the following years, the worldwide decrease in wildlife fur use led to a commercial crisis in the coypu industry. Attempts were made to establish new markets for coypu fur and meat; however, they were unsuccessful to revert the steady fall in demand. Consequently, the coypu breeding industry ceased in Mar Chiquita in 2016.

Capybara (*Hydrochoerus hydrochaeris*, Hydrochoerinae): Local name: *carpin-cho*. This largest living rodent is at present very common in the Dulce River wetlands, the western limit of the species geographical distribution (Parera 2002). There are no indications of the presence of the species in the area before the 1980s. It was first seen in 1984, and since then the population has grown considerably. The permanent presence of the capybara in the area was coincident with the Mar Chiquita highstand period and the flooding of the wetlands (Chap. 3). It is likely that the source of the expanding population came from the nearby Bajos Submeridionales in northern Santa Fe province (Parera 2002). Both wetlands are in close contact on the eastern border of the Mar Chiquita wetlands, particularly during wet years.

Chacoan marsh rat (*Holochilus brasiliensis*, Cricetidae)**:** Local name: *rata colorada*. A semiaquatic rodent species found in Argentina and Paraguay wetlands within the Gran Chaco ecoregion. The species is abundant across its distribution area (Parera 2002).

Tuco-tuco (*Ctenomys* spp., Ctenomyidae). Local name: *tuco-tuco*. The specific name of the species occurring in Mar Chiquita is uncertain since the taxonomy of this genus is still debated. The tuco-tucos live in excavated burrows, which maintain a relatively constant temperature and humidity level. They have many morphological adaptations to excavate the soil, including their body shape, reduced eyes, and strong limbs. Common in Mar Chiquita on dry sandy soil areas (Haro et al. 2006).

Maned wolf (*Chrysocyon brachyurus*, Canidae): Local name: *aguara guazu*. The maned wolf is the tallest of the wild canids; its long legs are likely an adaptation to the tall grasslands of its native habitat. In Argentina, this species prefers open seasonally flooded grasslands, including human-disturbed areas such as cattle ranches (Soler et al. 2015). In Mar Chiquita, the maned wolf reaches the southernmost part of its present distribution range, although in the past it was found farther south in Argentina (Queirolo et al. 2011). The maned wolf has been rare in Mar Chiquita according to historical records, with considerable population size variations over time and a growing trend in recent years. The species was not found in the area during the 1973–1980 period (despite intensive searching) and started to be recorded again in 1980 in the Dulce River wetlands (Miatello and Cobos 2008). Resighting of the maned wolf followed large-scale flooding of the Dulce River wetlands (Chap. 10). The presence of the maned wolf became more frequent in Mar Chiquita during the high water period. Sightings increased during the 1990s when the species expanded to the eastern and southern coasts of the lake. Since then, the maned wolf population has remained stable.

Similar to the situation of the capybara, the observed pattern of presence-absence of the maned wolf in Mar Chiquita suggests the possibility that the local population may have suffered extinctions during dry periods, with further recolonization during wet periods by migrants from the nearby Bajos Submeridionales wetlands, where there is a small, apparently stable population of the maned wolf (Soler et al. 2015).

Neotropical river otter (*Lontra longicaudis*, Mustelidae): Local name: *lobito de rio*. This species depends on a water environment with plenty of riparian vegetation. The distribution range extends into Uruguay, Paraguay, and across northern Argentina. In Argentina, due to excessive hunting in the 1970s, the otter populations became very low. Once they received full protection in 1983, their populations recovered rapidly. However, water pollution and habitat destruction through ranching and mining continue affecting this species (Parera 2002). The presence of the species in the Segundo River delta was first reported in 1904 and in Mar Chiquita in 1986. Since then, a small permanent population has been regularly observed in the same area (Haro et al. 2006).

Puma (*Puma concolor*, Felidae). Local name: *puma*. A large felid native to the Americas. Although no precise population estimations are available, reports from the local people indicate that this species is frequently seen in Mar Chiquita area, particularly in the dry woody vegetation that surrounds the flooding grasslands (Haro et al. 2006).

Jaguarundi (*Herpailurus yagouaroundi*, Felidae). Local name: *gato eira, yagua-rundi*. A small wild cat native to southern North America and South America. Its habitat is lowland brush and grasslands areas close to a source of running water. Frequently seen in the Mar Chiquita area (Haro et al. 2006).

Pampas fox (*Lycalopex gymnocercus*, Canidae). Local name: *zorro pampa*. The Pampas fox is a typical inhabitant of the Pampas grasslands. Common in open habitats and tall grass plains, also seen in open woodlands and in modified habitats, such as grazed pastures and croplands. Common in Mar Chiquita (Haro et al. 2006).

Southern tamandua anteater (*Tamandua tetradactyla*). Local name: *oso melero*. A medium-sized anteater that can climb trees. Feeds on ants and termites, using its extremely strong forelimbs to rip open nests and their elongated snouts and rounded tongues (up to 40 cm in length) to lick up the insects. The distribution area of the tamandua extends from Venezuela to northern Argentina. The presence of the tamandua in Mar Chiquita was suspected, particularly since the beginning of the twenty-first century due to a growing number of sights close to the Mar Chiquita area that expanded the known southern distribution range (Torres et al. 2009; Torres and Tamburini 2018). The presence of tamandua in Mar Chiquita was confirmed in 2016 with the finding of one road kill on the southern limit of the reserve, which adds support to the possibility of ongoing range expansion of the tamandua.

8.3 Unconfirmed Species

Crab-eating raccoon (*Procyon cancrivorus*, Procyonidae). Local name: *mayuato, osito lavador*. Native to marshy and jungle areas of Central and South America, rare in most of its range. The crab-eating raccoon is solitary and nocturnal, a good swimmer, almost always found near streams, lakes, and rivers. According to the species distribution range proposed by (Parera 2002), this species may occur in the Dulce River wetlands area. Precisely in this area (near Villa Candelaria), tracks were observed in 2001 (Haro et al. 2006).

Greater bulldog bat, fisherman bat (*Noctilio leporinus*, Noctilionidae). Local name: *murciélago pescador grande*. A fishing bat native to Latin America. Although not recorded in the Dulce River wetland, the reported distribution range includes the Dulce River basin down to close to the northern limit of the Mar Chiquita wetlands (Parera 2002).

Crab-eating fox, Savannah fox (*Cerdocyon thous*, Canidae). Local name: *zorro de monte*. Its presence in the Dulce River wetlands is very likely related to its habitat, which includes savannas and flooded grasslands, and to its known distribution range, which reaches the northwestern limit of Mar Chiquita area (Parera 2002; Torres and Tamburini 2018).

Grassland cat (*Leopardus pajeros*, Felidae). Local name: *gato de los pajonales*. The southern border of the species distribution range includes the Mar Chiquita area (Parera 2002). The Dulce River wetlands are, in principle, a suitable habitat for the species, given its preference for open areas, including woodland and grasslands.

Giant anteater (*Myrmecophaga tridactyla*, Myrmecophagidae). Common name: *oso hormiguero*. This species occurs in Central and South America. The species is present in multiple habitats, including grassland and rainforest. It feeds primarily on ants and termites. It is listed as vulnerable by the International Union for Conservation of Nature (UICN). The open grasslands of the Dulce River wetlands appear to be a suitable habitat for the species. There are no museum specimens from this region. However, the anteater was mentioned as occurring in the Dulce River wetlands (Kanter 1932). According to Parera (2002), in the past it probably had a southerner distribution, reaching 31°S.

8.4 Locally Extinct Species

Jaguar (*Panthera onca*, Felidae). Local name: *jaguar*. The jaguar is the most abundant cat species in the Americas. This cat had a wide distribution in Argentina, which is rapidly declining and is currently highly reduced. In Mar Chiquita it was hunted with some frequency between 1910 and 1920 in the Dulce River wetlands. The last individual of the region would have been hunted in 1958 (Haro et al. 2006).

Pampas deer (*Ozotoceros bezoarticus*, Cervidae). Local name: *ciervo de las Pampas*. According to the information gathered from local people, the species was present in the grasslands of Dulce River wetlands until the 1950s. The last individuals would have been sighted between 1957 and 1959 (Haro et al. 2006). A small population of the Pampas deer still remains in the flooded grasslands of the Bajos Submeridionales lowlands, close to the eastern border of the Dulce River wetlands (González et al. 2016; Pautasso et al. 2002). This population could be considered a potential source for the recolonization of the Mar Chiquita grasslands through the creation of corridors of suitable vegetation.

Guanaco (*Lama guanicoe*, Camelidae). Local name: *guanaco*. This species has not been mentioned for the region even in a detailed report on the Mar Chiquita local geography and fauna (Kanter 1932). It is important to notice, however, that this author includes in his map a toponym to the east of the Dulce River named "*El Guanaco*" suggesting that the species might have been present before that year. The possible occurrence of guanacos in Mar Chiquita in the past is supported by the fact that the guanaco was common and still occurs in the Salinas Grandes mudflats, about 150 km west from *El Guanaco*.

References

Curto E, Castellino R (2006) In: Bucher E (ed) Bañados del río Dulce y laguna Mar Chiquita (Córdoba, Argentina). Academia Nacional de Ciencias (Córdoba, Argentina), Córdoba, pp 285–293

Garcia Mata R (1973) Apuntes históricos sobre legislación argentina relacionada con la conservación de la nutria. Consejo de Investigaciones de la Universidad Nacional de Rosario, Argentina, Córdoba, pp 32–67

González S, Jackson III JJ, Merino ML (2016) *Ozotoceros bezoarticus*, pp. https://doi.org/10.2305/IUCN.UK.2016-2301.RLTS.T15803A22160030.en, IUCN

Griva E (1973) La Nutria: historia de su conservación y explotación Miramar, Mar Chiquita, Consejo de Investigaciones Científicas. Universidad Nacional de Rosario, Argentina

Haro JG, Michelutti P, Torres RM, Molli AF, Bucher EH (2006) In: Bucher EH (ed) Bañados del Rio Dulce y Laguna Mar Chiquita (Córdoba, Argentina). Academia Nacional de Ciencias (Córdoba, Argentina), Córdoba, pp 277–283

Kanter H (1932) La cuenca cerrada de la Mar Chiquita en el norte de la Argentina. Bol Acad Nac Cienc Córdoba (Argent) 22:285–232

Lozano P (1754) Historia de la conquista del Paraguay, Río de la Plata y Tucumán. Imprenta Popular (reimpression 1873), Buenos Airs

Miatello R, Cobos V (2008) Nuevos aportes sobre la distribución del aguará guazú *(Chrysocyon brachyurus*, Carnivora: Canidae) en las provincias de Córdoba y Santiago del Estero, Argentina. Mastozoologia Neotropical 15(2):209–213

Parera A (2002) Los mamíferos de la Argentina y la región austral de Sudamérica. El Ateneo, Buenos Aires

Pautasso A, Peña M, Mastropaolo J, Moggia L (2002) Distribución y conservación del venado de las pampas (*Ozotocerus bezoarticus leucogaster*) en el norte de Santa Fe, Argentina. Mastozoologia Neotropical 9:64–49

Queirolo D, Moreira J, Soler G, Emmons L, FH R, Pautasso A, Cartes A, Salvatori V (2011) Historical and current range of the near threatened maned wolf *Chrysocyon brachyurus* in South America. Oryx 45:2296–2303

Soler L, Orozco M, Caruso N, Gonzalez P, Pautasso A (2015) In: Orozco M, Gonzalez P, Soler G (eds) El aguará guazú *Chrysocyon brachyurus* en la Argentina. Lecciones aprendidas y recomendaciones para su conservación. Fundación de Historia Natural Félix de Azara, Buenos Aires

Torres R, Monguillot J, Bruno G, Michelutti P, Ponce A (2009) Ampliación del límite austral de la distribución del oso melero (*Tamandua tetradactyla*) en la Argentina Nótulas faunísticas Segunda Serie, 39, 1–5

Torres R, Tamburini, D (2018) Mamíferos de Córdoba. Universidad Nacional de Córdoba, Córdoba

Chapter 9
Mosquitos

9.1 Introduction

Mosquitoes are difficult-to-ignore components of wetland environments. Their high diversity and the large population size they may reach influence the ecosystem dynamics in very significant ways. First, as a component of the food chain during the larval stage in aquatic environments, and second, in the adult stage, as one of the most significant hematophagous organisms in wetlands, with the capacity of severely impacting animal and human health in terms of blood consumption and disease transmission.

In Mar Chiquita, mosquitos are considered a nuisance by the local population because of their negative impact on humans and domestic animals, as well as their potential role in terms of disease transmission (Bucher 2006). In addition, mosquitos also affect the Mar Chiquita wetland indirectly, first due to their negative effect on wetland acceptability by humans, and second as a consequence of the harmful impact of the insecticides used in mosquito control campaigns.

This chapter summarizes the existing information on mosquitos in Mar Chiquita, including the species list and a detailed account of *Aedes albifasciatus*, the most problematic species in terms of the large population outbreaks it may develop and the negative impacts generated by disease transmission on humans and domestic animals.

9.2 Mosquito Species Diversity in Mar Chiquita

To date, 33 mosquito species have been identified in Mar Chiquita. The present list of confirmed species is based on Bucher (2006) and Linares et al. (2016).

1. *Aedeomyia squamipennis*
2. *Aedes (Ochlerotatus) albifasciatus*
3. *Anopheles albitarsis*

© Springer Nature Switzerland AG 2019
E. H. Bucher, *The Mar Chiquita Salt Lake (Córdoba, Argentina)*,
https://doi.org/10.1007/978-3-030-15812-5_9

4. *Culex acharistus*
5. *Culex bidens*
6. *Culex brethesi*
7. *Culex chidesteri*
8. *Culex dolosus*
9. *Culex educator*
10. *Culex interfor*
11. *Culex maxi*
12. *Culex pipiens*
13. *Culex saltanensis*
14. *Haemagogus spegazzinii*
15. *Mansonia humeralis*
16. *Mansonia indubitans*
17. *Mansonia titillans*
18. *Ochlerotatus scapularis*
19. *Ochlerotatus stigmaticus*
20. *Psorophora albigenu varipes*
21. *Psorophora ciliata*
22. *Psorophora confinnis*
23. *Psorophora cyanescens*
24. *Psorophora dimidiata*
25. *Psorophora discrucians*
26. *Psorophora ferox*
27. *Psorophora holmbergi*
28. *Psorophora pallescens*
29. *Psorophora paulli*
30. *Psorophora varinervis*
31. *Uranotaenia apicalis*
32. *Uranotaenia lowii*
33. *Uranotaenia nataliae*

9.3   The Mosquito *Aedes albifasciatus*

Of all the mosquito species present in Mar Chiquita, *Aedes albifasciatus* is the most significant, given the enormous numbers it may reach and the problems they cause. In some years the population may be high enough to affect the local economy, particularly tourism, and to interfere with livestock raising, particularly cattle, affecting meat and milk production. In times when the mosquito population peaks, milk production in the Mar Chiquita region may decrease in about 22% (Bucher 2006).

Tourism is also severely affected. In the past, before the 1970s, control campaigns using aerial spray of insecticides were frequently required (Bucher 2006). There is also agreement in considering the species as potential dispersers of most of the virus strains present in Argentina, particularly the Western equine encephalitis (WEE) (Avilés et al. 1992; Sabattini et al. 1998; Ludueña Almeida et al. 2004).

9.3.1 Distribution and Habitat Preferences

This species is widely distributed in the Southern Cone of South America (Bolivia, southern Brazil, Chile, Uruguay, and Argentina) and is common in saline wetlands. In Mar Chiquita, this mosquito is common throughout the whole wetland, particularly in areas with a vegetation physiognomy known as halophytic scrub (Chap. 10). In these areas, the vegetation is characterized by a sparse cover of salt-adapted, succulent plant species. Dominant species include *Salicornia ambigua*, *Sesuvium portulacastrum*, and *Atriplex cordobensis* (Fig. 9.1) (Bachmann and Casal 1963).

9.3.2 Life Cycle

The information presented in this section summarizes the results of a research project developed in Mar Chiquita that resulted in significant advances in the knowledge of the species ecology (Ludueña Almeida and Gorla 1995; Gleiser et al. 1997, 2000, 2002; Ludueña Almeida et al. 2004).

Breeding *Aedes albifasciatus* females lay drought-resistant eggs on damp soil. There the eggs can remain in quiescence condition for at least a year and become

Fig. 9.1 The plant formation halophytic scrubland in the Dulce River Wetlands, the habitat most favorable for *Aedes albifasciatus*. *Salicornia ambigua* and *Sesuvium portulacastrum* are the dominant shrubs in the image

viable and hatch after the first rain that keeps them covered with water for at least 24 h. Under suitable temperature conditions, adults emerge synchronously in about 8 days. The simultaneous emergence of enormous clouds of mosquitos within a short period is a frequent event in Mar Chiquita, which causes serious problems to the local population, as mentioned previously.

Population Dynamics The *Aedes albifasciatus* population peaks during the rainy season (October–March), starting with the first rain events of sufficient intensity, although adults and immature stages may be found during winter. Females may live up to 50 days (Ludueña Almeida and Gorla 1995). The size of the mosquito cohort reaching the adult stage in a given area depends on the area flooded by rain and on whether or not the temporary pools remain long enough to allow the larvae to develop into adults.

The studies in Mar Chiquita showed a significant correlation between adult *Aedes albifasciatus* abundance and rainfall accumulated during the previous week, indicating that these variables could predict the emergence of adults within this time range. Moreover, Gleiser et al. (1997) showed that the presence or absence of larvae in the field could be predicted from temporal variations of the normalized vegetation index (NDVI) provided by low-resolution satellite images (Gleiser et al. 1997).

Research in Mar Chiquita also showed that the spatial distribution of population abundance seems to be influenced by local factors, considering that the breeding cycle at sampling sites located about 10 km was not synchronous. A possible key local factor could be the spatial heterogeneity of summer rain events in the area, in most cases originated from convective cumulus-nimbus clouds (Bucher et al. 2006).

9.3.3   Environmental Factors that Control Mosquito Outbreaks

Events of mosquito populations reaching very high levels were much commoner before the late 1970s when rainfall was much lower than in the subsequent years until present, and the lake level was also lower (Chap. 3). During this lowstand period, mosquito control campaigns using aerial spray of insecticides were necessary almost every year.

Surprisingly, higher precipitation during the highstands did not correlate with mosquito abundance, and in fact problems with mosquito surges showed a decreasing trend. Unfortunately, no reliable knowledge exists to explain year-to-year variations in mosquito abundance, which would be essential for more informed management strategies.

One interesting hypothesis that could be applied to the Mar Chiquita case has been proposed by Chase and Knight (2003). In summary, they hypothesize that mosquitos would show population outbreaks after irregular drought years. Specifically, it is suggested that in wetlands that never dry (permanent), and even in those that dry yearly (temporary), competitors and predators that are well adapted to predictable drying limit mosquito abundance.

On the contrary, in wetlands that dry only during semipermanent, unpredictable drought years, mosquito predators and competitors are eliminated and must recolonize following a drought, allowing mosquitos, which have a more rapid population dynamics, to achieve population outbreaks in a very short time. This hypothesis is consistent with the marked differences in mosquito abundance observed in Mar Chiquita, considering that during the lowstands, rainfall was much lower and the dry periods longer and more irregular (Chap. 3).

References

Avilés G, Sabattini MS et al (1992) Transmission of western equine encephalomyelitis virus by Argentine *Aedes albifasciatus* (Diptera: Culicidae). J Med Entomol 29:850–853

Bachmann AO, Casal OH (1963) Mosquitos argentinos que crian en aguas salobres y salinas. Rev Soc Entomológica Argent 25:21–27

Bucher EH (2006) In: Bucher EH (ed) Mosquitos de Mar Chiquita. Bañados del Rio Dulce y Laguna Mar Chiquita. Academia Nacional de Ciencias, (Córdoba, Argentina), Córdoba, pp 295–299

Bucher EH et al (2006) Síntesis geografica. In: Bucher EH (ed) Bañados del rio Dulce y laguna Mar Chiquita (Córdoba, Argentina). Academia Nacional de Ciencias (Córdoba, Argentina), Córdoba, pp 15–27

Chase JM, Knight TM (2003) Drought-induced mosquito outbreaks in wetlands. Ecol Lett 6(11):1017–1024

Gleiser RM et al (1997) Monitoring the abundance of *Aedes (Ochlerotatus) albifasciatus* (Macquart 1838) (Diptera: Culicidae) to the south of Mar Chiquita Lake, central Argentina, with the aid of remote sensing. Ann Trop Med Parasitol 91:917–926

Gleiser RM et al (2000) Population dynamics of *Aedes albifasciatus* (Diptera: Culicidae) South of Mar Chiquita Lake, Central Argentina. J Med Entomol 37(1):21–25

Gleiser RM et al (2002) Spatial pattern of abundance of the mosquito, *Ochlerotatus albifasciatus*, in relation to habitat characteristics. Med Vet Entomol 4:364–371

Linares M et al (2016) New mosquito records (Diptera: Culicidae) from northwestern Argentina. Check List 12(4):1944

Ludueña Almeida F, Gorla DE (1995) The biology of *Aedes (Ochlerotatus) albifasciatus* Macquart 1838 (Diptera: Culicidae) in central Argentina. Mem Inst Oswaldo Cruz, Rio de Janeiro 90:463–468

Ludueña Almeida F et al (2004) Culicidae (Diptera) del arco sur de la Laguna de Mar Chiquita (Córdoba, Argentina) y su importancia sanitaria. Rev Soc Entomológica Argent 63:25–28

Sabattini MS et al (1998) Historical, epidemiological and ecological aspects of Arboviruses in Argentina: Flaviviridae, Bunyaviridae and Rhabdoviri-dae. In: da Rosa ARAT, Vasconcelos PFC, Travassos da Rosa JFS (eds) An overview of Arbovirology in Brazil and neighbouring countries. Instituto Evandro Chagas, Belem, pp 135–153

Chapter 10
Dulce River Wetland

10.1 Introduction

The South American Chaco ecoregion was originally rich in wetlands, including freshwater and saline wetlands (Canevari et al. 2001). Unfortunately, the wetlands and grasslands of the region are disappearing rapidly due to deforestation, agriculture expansion, and water diversion (Grau et al. 2014). Among the remaining areas, the Rio Dulce wetland and its grasslands in the lower floodplain of the Dulce River Mar Chiquita Lake is considered one of the few remaining large-sized Chaco saline wetlands (Fig. 10.1).

In addition, Mar Chiquita holds an exceptional biodiversity richness, still mostly unaltered. These valuable characteristics have justified the inclusion of Mar Chiquita among the 20 high-priority areas for the conservation of Nearctic migratory birds in the southern cone grasslands of South America (Di Giacomo and Parera 2008). The area has remained isolated, underpopulated, and unknown until recently. The present knowledge about the area is summarized in the following sections.

10.2 Vegetation and Landscape

The Dulce River wetland is a typically flooded savanna, with a heterogeneous and complex landscape that combines the Dulce River flow, temporary and permanent ponds, extended grasslands, halophytic scrubs, and elevated areas with woody vegetation. The dominant plant formations in the Rio Dulce wetland are briefly described here based on Kanteer (1932), Sayago (1969), and Menghi (2006) (Fig. 10.2).

Chaco Forest The climax dry woodland that grows in the highland that surrounds the Mar Chiquita floodplain.

© Springer Nature Switzerland AG 2019
E. H. Bucher, *The Mar Chiquita Salt Lake (Córdoba, Argentina)*,
https://doi.org/10.1007/978-3-030-15812-5_10

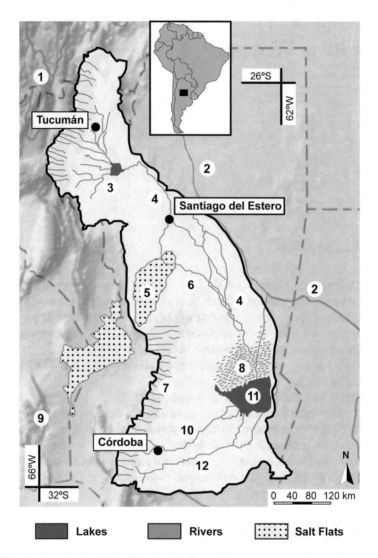

Fig. 10.1 Drainage basin of Mar Chiquita. (1) Sierras Pampeanas range (Aconquija mountains) (2) Salado River, (3) Rio Hondo dam, (4) Dulce River, (5) Ambargasta salt flats, (6) Saladillo branch of the Dulce River, (7) Western dispersed creeks, (8) Rio Dulce wetland, (9) Sierras Pampeanas range (Córdoba mountains) de Córdoba mountains, (10) Primero River, (11) Mar Chiquita Lake, (12) Segundo River. Dotted lines indicate provincial limits

Transition Forest and Shrubland (Fig.10.2(1)) A shrub steppe develops at a higher level and more distant from the river; it is characterized by a mixture of grasses and shrubs, with dominance of succulent shrubs, tall cactus, and isolated trees. The water table is generally close to the soil surface. A gradual change toward dominance of tree species is observed with increasing terrain elevation. As the

Fig. 10.2 Main vegetation formations on the Dulce River wetland. (1) Transition forest and shrubland, (2) halophytic shrubland, (3) spartina grassland, (4) riverine prairie, (5) riparian vegetation

altitude of the terrain and distance to the river increase, this vegetation is gradually replaced with the typical Chaco forest (Sayago 1969).

Halophytic Shrubland (Fig. 10.2(2)) This vegetation is characterized by a monotonous halophytic steppe with areas of open soil covered with salt crusts. Dominant species include succulent shrubs, including *Salicornia ambigua*, *Atriplex lorentzii*, and *Allenrolfea patagonica*. The area is flooded occasionally and by short periods, and the groundwater is very close to the soil surface (see also Fig. 10.3).

Spartina Grassland (Fig. 10.2(3)) Medium-height grassland community, dominated by grass species of the genus *Spartina*, particularly *S. argentinensis*, which are adapted to floods of lower frequency and shorter duration than the riverine

Fig. 10.3 Aerial view of
the Dulce River wetland
showing the contact area
between the halophytic
scrubland (isolated shrub
patches on open salt crust
soil) and the spartina
grassland (uniform green
cover). Very small
differences in soil level
determine variations in
water drainage and salinity
that control vegetation
cover. (From Bucher 2006)

prairies. These are grasses of lower forage value, although their nutritional content
is higher in the resprouts that grow after fires; for this reason, farmers regularly burn
them (see also Fig. 10.3).

Riverine Prairie (Fig. 10.2(4)) Close to the river, dominated by short grasses of
very high forage value, which become flooded almost every year at the end of the
rainy season (April–June).

Riparian Vegetation (Fig. 10.2(5)) The natural embankments along the braided
river branches in the wetland are usually dominated by the shrub *Baccharis salici-
folia* ("*chilca*") and the tree *Prosopis ruscifolia* ("*vinal*"), with scattered *Sapium
haematospermum* ("*lecherón*") trees, depending on topography and soil
characteristics.

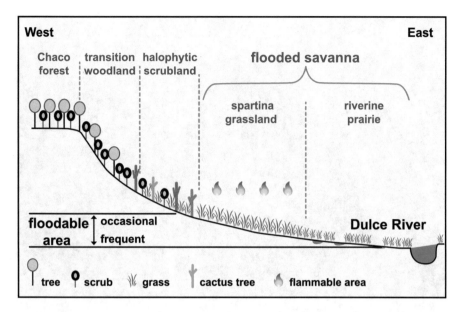

Fig. 10.4 Cross section of the floodplain of the lower Dulce River, from the river flow to the west. The succession of vegetation types with increasing terrain altitude and consequent decrease of annual flooding intensity and frequency is observed. (From Bucher 2006)

10.2.1   Factors Controlling the Distribution of Plant Formations

The distribution of the different plant formations in the floodplain depends mainly on soil salinity and proximity of groundwater and also on two fundamental annual pulses: flooding and fire. As a result, plant formations appear arranged in parallel strips along the river course, each one controlled by an increasing gradient of terrain elevation starting from the Rio Dulce baseline level. Elevation strongly conditions the plant's environment by determining whether the terrain will or will not be reached by the annual floods and, if so, with what frequency (Fig. 10.4).

10.3   Flooding and Fire as Ecosystem Modulators

Both environmental pulses are fundamental for the maintenance of the wetlands. Annual flooding is originated by the overflow of the Dulce River occurring by the end of the rainy season, i.e., approximately between March and July (Fig. 10.5). The flooded area moves slowly from the northern part of the Dulce River floodplain valley downstream toward the Mar Chiquita Lake. Water moves in the form of a shallow, widely expanded sheet flow called "a moving lake" by Kanter (1932). The size of the flooded area depends on the amount of water overflowing from the riverbed,

Fig. 10.5   Flood expanding from the Dulce River into the Rio Dulce Wetland. (From Bucher 2006)

ranging from null in years of scarce rainfall to an area as vast as over 4,000 km^2 in years of heavy rains in the upper and middle basin of the Dulce River (Chap. 3). Annual floods also govern other important geological and ecological processes: (a) modeling of the hydrological network via sediment deposition and removal, (b) washing out of salts from the soil, (c) deposition of sediments and nutrients to the flooded soils, and (d) recharging of the dispersed ponds that are not directly irrigated by the Dulce River, which increases the availability of freshwater aquatic habitat.

When waters recede and the cold and dry season starts, most of the spartina biomass dries up and becomes easily flammable. At this moment, fires may start easily due to lightning or human intervention (Fig. 10.6). Wildfires are common and frequent events in Mar Chiquita as in most flooded savannas worldwide (Whelan 1995). Fire prevents the expansion of woody vegetation since repeated fires are beneficial to grasses but with adverse effects on shrubs and trees. Hence, fire suppression (e.g., when fuel from grasslands is removed by overgrazing) leads to a successional process that leads to shrub and tree encroachment of the grasslands (Huntley and Walker 1982; Whelan 1995).

Fig. 10.6 Annual fires in the spartina grassland formation. (From Bucher 2006)

10.4 Biodiversity

The list of dominant plants recorded in the area is presented in Table 10.1. Information of the fauna of the Dulce River wetland is given in Chaps. 5 (Fish), 6 (Amphibians and Reptiles), 7 (Birds), 8 (Mammals), and 9 (Mosquitos).

10.5 Conservation and Sustainable Use

Wetland grasslands are disappearing around the world. The South American Chaco ecoregion, originally rich in wetlands, do not escape to this trend. Saline wetlands and grasslands are well represented in the region, particularly on the Western Chaco of Argentina. Among them, the Rio Dulce river grasslands are particularly important, considering their large area and its protected status in terms of a provincial reserve and a Ramsar site of international importance.

Table 10.1 Common plant species according to plant formations in the Rio Dulce wetland. (Refer to Fig. 10.4). From Menghi (2006)

Species	Flooding savanna	Halophytic shrubland	Transition woodland	Chaco forest
Elevation	72.5–74.0	74–76	76–94	>94
Scirpus californicus	x			
Typha latifolia	x			
Scirpus americanus	x			
Baccharis juncea	x			
Chenopodium macrospermum	x	x		
Distichlis spicata	x	x		
Malvella leprosa	x			
Melilotus indicus	x			
Frankenia pulverulenta	x			
Ambrosia tenuifolia	x			
Aster squamatus	x			
Polypogon monspeliensis	x	x		
Diplachne univervia	x			
Spergularia levis	x			
Eleocharis macrostachya	x			
Polygonum aviculare	x		x	
Rumex obtusifolius	x			
Lepidium bonariense	x			
Cynodon dactylon	x	x	x	
Sesuvium portulacastrum	x		x	
Spartina argentinensis	x	x	x	
Salicornia ambigua V	x	x	x	
Heliotropium curassavicum	x			
Cressa truxillensis	x	x	x	
Petunia parviflora	x	x	x	
Boopis anthemoides	x	x	x	
Sphaeralcea mineata	x	x	x	x
Sporobolus phleoides	x	x	x	
Phyla canescens	x	x		
Baccharis salicifolia		x	x	
Bothriochloa saccharoides		x	x	
Holocheilus hieracioides		x	x	
Allenrolfea patagonica		x	x	
Lepidium aletes		x		
Lycium americanus		x	x	
Parietaria debilis		x	x	x
Atriplex cordobensis		x	x	
Lycium infaustum		x	x	

(continued)

Table 10.1 (continued)

Species	Flooding savanna	Halophytic shrubland	Transition woodland	Chaco forest
Harrisia pomanica		x	x	
Lipipia salsa		x	x	x
Cyclolepis genistoides		x	x	
Allenrolfea vaginata		x	x	
Sporobolus pyramidalis	x	x	x	
Prosopis reptans	x	x	x	
Aloysia gratissima		x	x	
Hybanthus parviflorus		x	x	
Spergularia marina		x	x	
Mollugo verticillata		x	x	
Partenium hysterophorus			x	
Larrea divaricata			x	
Leptoglossis linifolia		x	x	
Cleistocactus baumannii		x	x	x
Lycium chilense		x	x	
Opuntia quimilo	x	x	x	
Schizachyrium microstachyum	x	x	x	
Trichloris crinita	x	x	x	x
Plantago myosuros	x	x	x	
Ruellia tweed	x	x		
Heimia salicifolia	x	x		x
Euphorbia serpens	x	x		
Eragrostis lugens	x	x	x	
Elionurus viridulus		x		
Cardus sp.		x	x	x
Geoffroea decorticans		x	x	x
Grabowskia duplicata		x		x
Maytenus vitis-idaea		x	x	x
Setaria spp.		x	x	x
Prosopis nigra			x	x
Relbunium bigeminum		x	x	x
Clematis sp.		x	x	x
Cortesia cuneifolia		x	x	x
Pappophorum mucronulatum	x	x	x	x
Gnaphalium sp.		x	x	
Stipa eriostachya			x	x
Prosopis sericantha			x	x
Cereus coryne			x	x
Bothriochloa laguroides	x	x	x	x

(continued)

Table 10.1   (continued)

Species	Flooding savanna	Halophytic shrubland	Transition woodland	Chaco forest
Solanum pigmaeum			x	
Ephedra triandra			x	
Lycium tenuispinosum			x	x
Gamochaeta sp.		x		x
Passiflora mooreana	x	x		
Bumelia obtusifolia				x
Maytenus spinosa				x
Ziziphus mistol			x	x
Solanum glaucum				x
Celtis spinosa				x
Condalia microphylla			x	x
Jodina rhombifolia			x	x
Prosopis alba				x
Capsicum chacoense			x	x
Capparis atamisquea			x	x
Porliera microphylla			x	x
Aspidosperma quebracho-blanco		x	x	
Dichondra microcalyx				x
Oxalis sp.			x	x
Rhynchosia senna			x	x
Chaptalia nutans		x		x
Schinus polygamus		x	x	x

10.5.1   Human Influence on the Dulce River Wetland: A Historical Review

The available evidence indicates that before the arrival of the Europeans, the Dulce River wetland was inhabited by several native groups, probably nomads that visited the area temporarily for fishing or hunting. These groups used fire for cooking and hunting and even as a war weapon. As a result, they may have increased fire frequency with respect to fires ignited by natural causes, mostly lighting (Bucher et al. 2006a).

After the arrival of the Europeans, almost the entire Chaco region (including the Dulce River) remained under the control of the indigenous people for about three centuries, with occasional incursions of the Spanish colonizers and Jesuit missioners (Chap. 11). As a result, the Dulce River wetlands remained almost unaltered, given that the low agricultural potential of the soil and the frequent raids of hostile natives discouraged settlers from establishing in the region. Only in 1860, when the Argentine army dominated the entire Chaco, did effective occupation of the region start. Since then, a very portion of the European immigrants settled in the wetland region, composed of transhumant farmers and coypu (*Myocastor coypus*) hunters. Immigrants tended to settle around sites with freshwater availability, especially along the Dulce River and close to freshwater ponds fed by rainfalls and periodical floods of the Dulce River (Bucher et al. 2006b, Chap. 11).

Since 1860 and up to the end of the twentieth century, the dominant land use in the wetlands consisted of a subsistence economy based on extensive livestock production (mainly cattle and sheep) on the native flooded grasslands. Management consisted of transhumant pastoralism, favored by the lack of fencing in the entire area. During the low water periods (winter and spring), livestock grazed in the Dulce River prairies, and during the high water period (summer and autumn), the herds were taken to higher lands.

During the second half of the twentieth century, the Rio Hondo dam was built on the Dulce River, starting the large-scale regulation of the Dulce River. During the filling of the reservoir, started in 1972, the Dulce River flow decreased to the point that the water did not reach the Mar Chiquita Lake (Bucher et al. 2006b). The new dam allowed the expansion of irrigation projects mostly in the Santiago del Estero province, which reduced the water flow reaching Mar Chiquita below the quota established by the existing inter-provincial agreements between Santiago del Estero and Córdoba (Bucher et al. 2006a; Gallego 2012). More restrictions on water availability are to be expected since more dams and other hydraulic projects are in the planning by the Santiago del Estero province (Gallego 2012).

Starting in the early twenty-first century, new land-use practices were introduced in the region, due to the arrival of investors from outside the region, in some cases with property titles of dubious validity, through a process frequently referred to as "land grabbing" (Suliema 2018). These new stakeholders replaced the traditional extensive livestock production with an intensively managed system, based on the replacement of the native vegetation with cultivated pastures, and the extensive fencing of the area. This shift in land use continues growing at present, although there are indications that, besides its negative environmental impact, the economic sustainability of the new production system may be less profitable than expected, due to the high salinity of the soils and the high cost and scarcity of water needed for pasture irrigation and cattle breeding (Bucher 2016). This difficulty has been repeatedly observed in wetland grasslands worldwide (Homewood 2008; McGahey et al. 2014).

Key negative impacts of this new land-use system include (a) loss of the natural vegetation, (b) increasing fencing that difficults displacement of the wildlife, and (c) increasing pollution from agrochemicals. Moreover, extensive fencing and land subdivision makes transhumant livestock management unviable.

Intensive cattle raising and cultivation in the Rio Dulce wetland has the additional disadvantages of not only affecting negatively biodiversity conservation but also the value of the tourism generated by the natural attractions of the area. For example, fishing tourism in the Rio Dulce, a rapidly growing activity, may provide important benefits to local populations and the protected areas. Moreover, non-extractive, sustainable land uses such as tourism are among the activities recommended by the Ramsar Convention in the Ramsar sites (The Ramsar Convention 2013).

It is worth mentioning that the recent development of land-use conflicts in the Dulce River wetland is not an isolated problem restricted only to this particular wetland. The new tendency for large-scale investors to acquire pastoral rangelands for commercial crop farming and the resulting negative effects on the local population is now globally widespread, particularly in the African wetlands (Homewood 2008; Gomez Pinto de Abrieu et al. 2010; Suliema 2018) and also in the large Pantanal wetland in Brazil (Gomes Pinto de Abrieu et al. 2010).

10.5.2   Present Trends and Conservation Challenges

As described in the previous section, the Dulce River wetland is threatened by two processes at different landscape scales: first, the continued reduction in water flow and alteration of the natural flooding regime caused by upstream dams and irrigation developments and, second, drastic changes in land use.

Both factors are generating negative environmental impacts and also social conflicts, being particularly problematical in the protected area in the province of Córdoba. Unless these threats are recognized and adequate management measures implemented at both the river basin and local scales, the Rio Dulce wetland will continue under threat.

References

Bucher EH (ed.) (2006) Bañados del Rio Dulce y Laguna Mar Chiquita. Academia Nacional de Ciencias, Córdoba

Bucher E (2016) El futuro incierto de los humedales del Chaco: el caso de los Bañados del Río Dulce. Paraquaria Nat 4(2):11–18

Bucher E, Coria R, Curto E, Lima J (2006a) Conservación y uso sustentable. In: Bucher E (ed) Bañados del Rio Dulce y Laguna Mar Chiquita (Córdoba, Argentina). Academia Nacional de Ciencias (Córdoba, Argentina), Córdoba, pp 327–341

Bucher E, Marcellino A, Ferreyra C, Molli A (2006b) Historia del Poblamiento Humano. In: Bucher E (ed) Bañados del Rio Dulce y Laguna Mar Chiquita (Córdoba, Argentina). Academia Nacional de Ciencias (Córdoba, Argentina), Córdoba, pp 301–325

Canevari P, Davidson I, Blanco DE, Castro G, Bucher EH (eds) (2001) Wetlands of South America, An Agenda for Biodiversity Conservation and Policies Development. Wetlands International, Buenos Aires

Di Giacomo A, Parera A (2008) Veinte áreas prioritarias para la conservación de las aves migratorias en los pastizales del cono sur de Sudamérica. Aves Argentinas, Buenos Aires

Gallego A (2012) Santiago del Estero y el agua: crónica de una relación controvertida. Lucrecia Editorial, Santiago del Estero

Gomes Pinto de Abrieu U, McManus C, Santos SA (2010) Cattle ranching, conservation and transhumance in the Brazilian Pantanal. Pastoralism 1(1):99–114

Grau RN, Torres R, Gasparri IN, Blendinger PG, Marinaro S, Macchi L (2014) Natural grasslands in the Chaco. A neglected ecosystem under threat by agriculture expansion and forest-oriented conservation policies. J Arid Environ 123:40–46

Homewood K (2008) Ecology of African pastoralist societies. James Currey, Oxford

Huntley BJ, Walker BH (eds) (1982) Ecology of Tropical Savannas. Springer, Heidelberg

Kanter H (1932) a cuenca cerrada de la Mar Chiquita en el norte de la Argentina. Bol Acad Nac Cienc Córdoba (Argentina) 22:285–232

McGahey D, Davies J, Hagelberg N, Ouedraogo R (2014) Pastoralism and the Green Economy – a natural nexus? IUCN and UNEP, Nairobi

Menghi M (2006) Vegetation. In: Bucher EH (ed) Bañados del Rio Dulce y Laguna Mar Chiquita. Academia Nacional de Ciencias (Córdoba, Argentina), Córdoba, pp 1743–1189

Sayago M (1969) Studio fitogeográfico del norte de Córdoba. Bol Acad Nac Cienc Córdoba 46:123–427

Suliema HM (2018) Exploring the spatio-temporal processes of communal rangeland grabbing in Sudan. Pastoralism 8:14

The Ramsar Convention (2013) The Ramsar convention manual: a guide to the convention on wetlands (Ramsar, Iran, 1971), 6th edn. Ramsar Secretariat, Gland

Whelan RJ (1995) The ecology of fire. Cambridge University Press, Cambridge, UK

Chapter 11
History of Human Settlement

11.1 Introduction

There is widespread archaeological evidence of the presence of indigenous peoples in Mar Chiquita before the arrival of the Spanish conquerors, which may go as far in time as the first waves of humans coming to South America (Politis et al. 2016). Early references on the original peoples of Mar Chiquita (Aparicio 1942; Frenguelli and De Aparicio 1932; Oliva 1947) described the archaeological findings made on the southern coast of the Mar Chiquita Lake and in the Rio Dulce wetlands. Unfortunately, most of the archaeological sites found before the 1970s were lost due to the 1970 flooding of the Mar Chiquita Lake (Chap. 4). Since late in the twentieth century, additional field work has rendered the discovery of new sites, which are detailed in (Bonofiglio 2004; Bucher et al. 2006). Most of the collected material is kept in local museums (Museo de la Región de Ansenuza Aníbal Montes, in Miramar; Museo Histórico Municipal de La Para; and Museo Regional de Morteros (Bonofiglio 2004).

Among the oldest remains discovered in the region, an entire skeleton found near the city of Miramar on the southern coast of the lake in 1957 (known as the "man of Miramar") appears of primary interest. However, the available information is somewhat contradictory and requires confirmation. Initially, the age of the remains was estimated in about 12,000 years based on the strata where it was found (Montes 1960). However, further studies using more advanced techniques are needed to obtain more accurate age estimation.

Since late in the twentieth century, additional field work has rendered the discovery of new sites, which are detailed in Bonofiglio (2004), Bucher et al. (2006), and Ferreyra et al. (2013).

© Springer Nature Switzerland AG 2019
E. H. Bucher, *The Mar Chiquita Salt Lake (Córdoba, Argentina)*,
https://doi.org/10.1007/978-3-030-15812-5_11

11.2   First Arrivals: The Paleoamericans

The presence of the early hunter-gatherers known as "Paleoamericans" in Mar Chiquita needs confirmation. The possibility that humans were present in Mar Chiquita over 14,000 years ago cannot be discarded, taking into account that human remains dated ca. 14,000 years B.P were found in the Arroyo Seco 2 site, at about 800 km south of Mar Chiquita (Politis et al. 2016). The Arroyo Seco 2 finding adds to the growing list of American sites that indicate a human occupation earlier than the Clovis dispersal episode in the southern cone of South America, although posterior to the onset of the deglaciation of the Last Glacial Maximum (LGM) in North America (Politis et al. 2016). The more likely routes of dispersal of the first human groups coming from the north into Mar Chiquita were along the Dulce River or the Paraguay-Parana rivers (Rodriguez and Ceruti 1999).

11.3   Middle and Late Holocene Hunter-Gatherers

There is ample evidence of the presence of relatively advanced groups in Mar Chiquita at the initial stages of agriculture during the middle and late Holocene that were capable of producing relatively sophisticated stone and ceramic artifacts. Rodriguez and Ceruti (1999) assigned these groups to an archaeological entity called "Esperanza," a member of a cultural tradition typical of the central Argentine plains which occurred during a long period of about 1000 years (between 2000 and 1000 years BP), under a climate of generalized aridity that ends with the Humid Medieval Stage. The possibility that some small populations may have survived until the Europeans arrival in the region cannot be discarded (Rodriguez and Ceruti 1999)

Fig. 11.1 Archaeological material collected in Mar Chiquita from outside the region: stone from the western Pampean Ranges, (2) net-imprinted ceramics from the eastern Parana River, (3) bird-head ceramic from the Santiago del Estero plains, and (4) mollusk-shell necklace probably from the Pacific Ocean. (From Bucher et al. 2006)

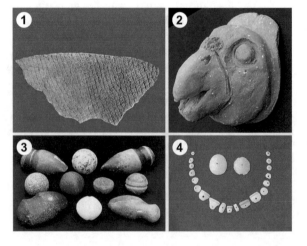

According to Del Rio et al. (2016), the Esperanza group was probably seminomadic, moving long distances throughout the year following a relatively fixed pattern that allowed an optimal use of the available resources. The more likely itinerary would include fishing along the Parana River and nearby wetlands in winter, moving to Mar Chiquita where hunting of the coypu rodent (*Myocastor coypus*) and birds was plentiful in spring and moving west to the mountains (Sierras de Córdoba) during summer to collect *algarrobo* tree fruits (*Prosopis alba*) (Fig. 11.1). See also Ceruti and González (2007) for an overview of the human adaptations to the wetland environments in central Argentina during the Late Holocene.

11.3.1   Morphological and Genetic Diversity

With regard to the human remains found in Mar Chiquita, recent evidence from studies based on two different methodological approaches (craniometric and molecular genetics) have rendered somewhat different results. Studies and the morphological analysis of other similar research results suggest a high morphological diversity among recent South American groups, which contrast the molecular studies that demonstrate a general loss of genetic diversity associated with increased distance from Africa. According to Marcelino and Colantonio (1993), craniometric measurements indicate a great diversity of the human groups that inhabited the Mar Chiquita region. Most of the remains recovered in the area, including the "Miramar man," belong to the Paleoamerican type, a classification term given to hunter-gatherers that used basic tools, who arrived and expanded in the Americas during the final glacial episodes of the late Pleistocene period.

The cranial characteristics of this group differ markedly from those of other specimens of agriculturalist-ceramist cultural identities that occupied the region in the late Holocene (such as the abovementioned Esperanza group) and differ from the neighboring Chaco ethnic groups (Bucher et al. 2006; Marcelino and Colantonio 1993).

These conclusions are not supported by more recent studies on the identification of the origin and lineage connections of the Mar Chiquita human remains using mitochondrial DNA. Results indicate a much more homogeneous genetic structure of all the remains that cannot sustain the population diversity suggested by the craniometric analysis (Nores et al. 2017).

These contradictory results from Mar Chiquita are another example of the many similar research results showing a high morphological diversity among recent South American groups, which contrast the molecular studies that demonstrate a general loss of genetic diversity associated with increased distance from Africa (Perez et al. 2009). In fact, this incongruence between phenotypic and genetic variance remains as an ongoing debate about the settlement and peopling of the Americas. A review of the possible factors involved in this apparent contradiction is presented in (Perez et al. 2009).

11.4   The Native Populations in Mar Chiquita at the Time of the Spanish Arrival

When the Spanish conquerors arrived in Córdoba and Santiago del Estero (early sixteenth century), the region surrounding the lower reach of Dulce River and the coasts of Mar Chiquita Lake were occupied by the *Sanavirones* ethnic group (also known as Yugitas by the Spaniards) (Serrano 1945; Berberian 1999).

The available information about *Sanavirones* is limited. They were hunter-gatherers as well as farmers, growing maize, beans, and squash. This highly mobile group kept permanent contact with neighboring ethnic groups (both in war and in peace times), particularly with the *Comechingones*, the ethnic group that populated the Sierras de Córdoba mountains to the west and south of Mar Chiquita (Berberian 1999; Montes 2008; Serrano 1945).

The influence of groups from the Parana River littoral was also detected, especially of the Goya-Malabrigo cultural entity, which is evidenced by frequent and relatively abundant remains of ceramic material from this region in the Mar Chiquita area (Del Rio et al. 2016). The area controlled by *Sanavirones* was also inhabited by smaller ethnic groups, which survived until the arrival of the Spaniards. Two of these groups, the *Malquesis* and *Quelosis*, lived in isolated tribes on the north of the Mar Chiquita Lake, on the Dulce River wetlands. The Jesuit priest Pedro Lozano (1754) reports that these groups "paid their tributes with coypus whose offspring were their only food source, drank brackish water and imitated all of the fauna nature in such a way that they seemed more abnormal birds from those lakes than living humans" (coypus is a very common rodent in the area, as described in Chap. 8). According to Rodriguez and Ceruti (1999), these groups might be the last descendants of the previously described Esperanza cultural entity.

11.5   The European Occupation

11.5.1   The Discovery

Mar Chiquita region was discovered in 1545 by the Spanish expedition led by Diego de Rojas (1543–1546) originated in Lima, Peru (Serrano 1945; Piossek 1995). This expedition set off in Lima, Peru, with the purpose of exploring and incorporating lands to the Spanish domain and establishing a communication route with the La Plata River and the Atlantic Ocean.

According to the cartography available in those days, it is very likely that the expeditionaries were assuming that the Dulce River was a direct tributary of the Parana River, which in turn would lead to the La Plata River, their final objective (Montes 1960).

With this goal, they advanced downstream of the Dulce River until they arrived in the totally unexpected, extended Dulce River wetlands. In the chronicles of the expedition, there are references of a group of soldiers that decided to advance to the wetland area in 1545, guided by a local native (expedition described as the "entry to the mudflats"), where they found numberless difficulties. Finally, they arrived in an open salt flat (probably the salt mudflats on the north coast of the Mar Chiquita Lake), from where they were forced to return (Piossek 1995). The chronicles also mention that the soldiers managed to survive due to a large number of eggs of water birds that they found in the area (probably flamingos) (Serrano 1945). At this point, the expedition abandoned the Dulce River flow and moved southward, taking another direction more to the west, following the trails frequently used by the *Sanavirones* to communicate with the *Comechingones* group in the Sierras de Cordoba Mountains (Montes 1960).

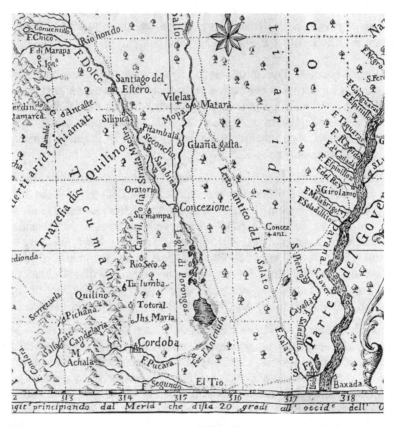

Fig. 11.2 Map of the Gran Chaco drawn by the Jesuit priest Joaquín Camargo Bazán, published in 1780 (Jolis 1972). In this map the Salado River does not flow into the Parana River. Instead, it appears joining the Dulce River and flowing into the Mar Chiquita Lake. The abandoned Salado River old riverbed is also indicated

11.5.2   Land Occupation

The Mar Chiquita region remained excluded from the mainstream land occupation process by Spaniards for a long time, practically until the end of the nineteenth century. This delay was influenced by three main factors: (1) the hostile territory made human occupation difficult, (2) the permanent threat posed by the indigenous people from the region, and (3) the fact that the main communication route between Peru and the La Plata River ran to the south and west of Mar Chiquita (Fig. 11.2). The details of this historical process are described in the following sections.

Sixteenth Century After being first discovered in 1543, the Mar Chiquita area remained almost unexplored and isolated for the rest of the century. At the end of the century, the Jesuit priests arrived in Cordoba. One of their main objectives was to explore the Gran Chaco region (which includes Mar Chiquita) and contact the native populations and attract them to communities under their control and support (named "missions" or "reductions"). They played a very significant role in the European expansion in the Gran Chaco region until their expulsion from South America in 1759 (Levillier 1931).

Seventeenth Century During this period, two contrasting processes appear as a key regarding the European occupation of the Mar Chiquita region: first, the expansion of trade between Lima in Peru and Buenos Aires on the Atlantic Coast and second the rapid adoption of the horse by several Chacoan native tribes, which significantly increased their mobility and offensive capability in their incursions into the Europeans settlements.

The expansion of trade and silver contraband through Cordoba led the Spanish crown to establish a tariff line, the "Aduana Seca," in Córdoba between 1622 and 1696 (Carrió de la Vandera 1908). As a result, smugglers resourced to alternative roads to avoid Cordoba across the Dulce River wetlands area and then heading north along the Dulce River up to Santiago del Estero, Tucuman, finally reaching Bolivia and Peru (Carrió de la Vandera 1908). This shortcut was scarcely used due to the adverse conditions of the trail that crossed marshy areas and saline mudflats, besides the risks of frequent attacks by indigenous people (Carrió de la Vandera 1908). Parts of the trail are still visible today, particularly close to Los Porongos lagoon.

Eighteenth Century At the beginning of the century, the Chacoan native groups, Abipones, Mocovíes, and Guaycurúes, were fully adapted to the use of the horse, which provided them with high mobility in their warfare against the Spanish colonizers. As a result, the frontier between the area dominated by Spanish and the region occupied by indigenous people underwent several setbacks, particularly along the eastern and southern borders of the area (Dobrizhoffer 1822; Punzi 1997). Hence, the frontier line became very vulnerable and collapsed on several occasions, exposing the region to continued raids that penetrated deeply in the area occupied by the European settlers (Dobrizhoffer 1822). During this century complete and accurate maps of the entire Chaco region, including Mar Chiquita, were produced by the Jesuit priests that had traveled extensively in the region. Of particular interest is the map of Joaquín Camargo Bazán in which it is documented the deviation of the Salado River that joined the Dulce River (Fig. 11.2).

Nineteenth Century During the first half of the century, Mar Chiquita continued under the native's control. Toward 1801, the last indigenous reductions created in the Gran Chaco by the Jesuits expulsed in 1767 were abandoned; as a consequence, Europeans were no longer present in the entire region. Consequently, the land occupied by Europeans was still frequently overrun by horse-riding groups. This situation aggravated still further due to the independence wars against Spain and the subsequent civil wars. The troops that were deployed in the frontiers of Mar Chiquita were removed, and most of the frontier line and the route between Santa Fe and Cordoba collapsed. This hostile panorama in the Gran Chaco only improved after the civil war was over, and the constitution of the country was definitively established in 1864. Shortly after that (1874), the Argentine army started a systematic occupation of the so-called Indian land in the Gran Chaco that led to the total occupation of the region early in the twentieth century. The Mar Chiquita region was secured by 1869, as the frontier line was displaced north of the region (Punzi 1997) (Figs. 11.3 and 11.4). The Mar Chiquita region was entirely secured by the

Fig. 11.3 Displacement of the frontier line between the territories controlled by the Spaniards and the Chaco native Americans. The dotted line corresponds to the frontier line hold until 1858. The solid line indicates the new line established in 1865, opening the Mar Chiquita to European settlement. The ruins Fuerte de la Costa was the closest fort to Mar Chiquita coast. (Data from Punzi 1997)

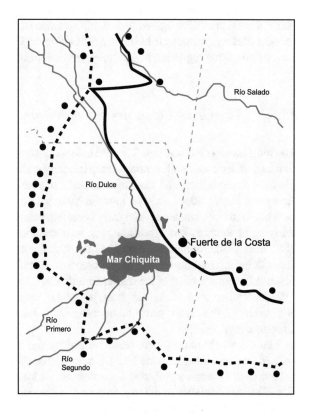

Fig. 11.4 The Fuerte de la Costa (Mar Chiquita coast fort). The closest fort to the lake along the 1865 frontier line against the raids of the Chaco natives. The fort ruins are still in place. (From Museo Regional de Morteros, Córdoba)

Europeans late in the nineteenth century, about 300 years after the first Spanish entry in the region. During that period, the area was excluded from the development process that was witnessed in other neighboring regions, remaining in the almost pristine condition regarding its landscape and wildlife.

11.5.3   The Great European Immigration

The long-lasting civil war that followed independence of Argentina in 1810 took the first half of the nineteenth century. Once it finally ended, the country entered a new phase of organization and development, which prioritized a policy of promoting European immigration. This large-scale human influx started during the 1860s, peaking in the late nineteenth and early twentieth centuries, and consisted mostly of Italian and Spanish immigrants. About four million Europeans landed between 1881 and 1930. Immigrants expanded rapidly, favored by the extensive railroad network built across the fertile Pampas plains of central Argentina. By the end of the nineteenth century, most of the region around Mar Chiquita was in private hands. In 1901, a detailed atlas of real estate in Argentina was published, in which the property titles of the lands surrounding Mar Chiquita were already included (De Chapeaurouge 1901).

The twentieth century was characterized by the rapid population growth and settlement consolidation in the Mar Chiquita region (Bucher et al. 2006). The original native dry woodland forest that surrounded the lake on these sectors was gradually eliminated following the expansion of agriculture, which intensified after the second half of the century (Cabido and Zak 1999).

Fig. 11.5 The coastal city of Miramar flooded in 1983 due to a marked increase in the Mar Chiquita Lake water level. (From Bucher et al. 2006)

Starting in the 1970s, two independent, somewhat contradictory events occurred almost simultaneously, which led to a turning point in the environmental situation of Mar Chiquita and therefore impacted deeply on the local population. First, the building of the Rio Hondo dam in Santiago del Estero together with the development of large agriculture irrigation projects led to a substantial reduction in the water flow (and the annual flooding regime) reaching Mar Chiquita. Second, a few years later, a rapid and intense increment in the rainfall regime throughout the whole Dulce River drainage basin led to a dramatic increase in the Mar Chiquita Lake, causing the largest expansion of Mar Chiquita in its entire geological history (Chaps. 2 and 3). As a result, about half of the only existing town on the lake shore (Miramar) was flooded and partly devastated (Fig. 11.5). Increased rainfall also accelerated a rapid expansion of agriculture around the whole Mar Chiquita wetland, which by the beginning of the twenty-first century had replaced the native dry woodland almost entirely (Cabido and Zak 1999).

Also in the Dulce River wetlands, there were significant changes that started at the end of the twentieth century, caused by two processes acting at different landscape scales. First, a reduction in the Dulce River water flow and increasing control of the natural flooding regime, due to the already mentioned water deviations in the upper sections of the river, and second, a rapid change in land use from transhumant pastoralism based on moving livestock according to the annual flooding to a breeding system based on extensive fencing and replacement of the native grasses by introduced pastures. These changes are generating negative environmental impacts and social conflicts, being particularly problematic in the protected area (Chap. 10) (Bucher 2016).

References

Aparicio F (1942) Arqueología de la laguna de Los Porongos. Relac Soc Argent Antropología 3:45–52

Berberian EE (1999) Nueva Historia de la Nación Argentina. In: De Marco M (ed) 1: La Argentina aborigen, conquista y colonización. Editorial Planeta, Buenos Aires, pp 135–158

Bonofiglio M (2004) Estudios iniciales en la cuenca baja de los ríos Suquía y Xanaes. Rev Museo Histórico Munic La Para 4:23–44

Bucher EH (2016) El futuro incierto de los humedales del Chaco: el caso de los bañados del Rio Dulce. OParaquarua Nat 4(2):11. 18

Bucher E, Marcellino A, Ferreyra C, Molli A (2006) In: Bucher E (ed) Bañados del Rio Dulce y Laguna Mar Chiquita (Córdoba, Argentina). Academia Nacional de Ciencias (Córdoba, Argentina), Córdoba, pp 301–325

Cabido M, Zak M (1999) Vegetación del norte de Córdoba, Secretaria de Agricultura, Ganaderia y Recursos Renovables de la provincia de Córdoba, Argentina, Córdoba, Argentina

Carrió de la Vandera AC (1908) El lazarillo de ciegos caminantes desde Buenos Aires hasta Lima (1773). Biblioteca de la Junta de Historia y Numismática Americana, Buenos Aires

Ceruti C, González MI (2007) Modos de vida vinculados con ambientes acuáticos del nordeste y pampa bonaerenses de Argentina. Relac Soc Argent Antropología 32:101–140

De Chapeaurouge C (1901) Atlas Catastral de la República Argentina. Oficina Cartográfica de Pablo Ludwig, Buenos Aires

Del Rio P et al (2016) Arqueología de los Bajos Submeridionales: sitios con hornos de tierra cocida en la localidad arqueológica Laguna La Blanca (La Criolla, Departamento San Justo, Provincia de Santa Fe). Rev Antropología Museo Entre Ríos 2(2):68–83

Dobrizhoffer M (1822) The Abipones, an equestrian people of Paraguay. John Murray, London

Ferreyra C et al (2013) Arqueologia del Mar de Ansenuza: actualizacón y nuevas investigaciones. Museo Histórico Municipal "La Para" (Córdoba, Argentina), La Para

Frenguelli J, De Aparicio F (1932) Excursión a la Laguna de Mar Chiquita (Provincia de Córdoba). Publicaciones del Museo Antropológico y Etnográfico de la Facultad de Filosofía y Letras. Univ Buenos Aires Ser A 2:121–132

Jolis J (1972) Ensayo sobre la historia natural del Gran Chaco. Universidad Nacional del Nordeste, Argentina. Instituto de Historia, Resistencia

Levillier R (1931) Nueva Crónica de la Conquista del Tucumán, vol II. Editorial, Buenos Aires, pp 1563–1573

Lozano P (1754) Historia de la conquista del Paraguay, Río de la Plata y Tucumán. Imprenta Popular (reimpression 1873), Buenos Aires

Marcelino A, Colantonio S (1993) Relaciones morfológicas de los aborígenes prehispánicos del territorio argentino VII: la Región Serrana de Córdoba. Rev Junta Provincial Hist Córdoba 8:113–135

Montes A (1960) El hombre fósil de Miramar. Revista de la Facultad de Ciencias Exactas, Físicas y Naturales, Universidad Nacional de Córdoba, Argentina. Ser Cienc Nat 21:1–29

Montes A (2008) Indígenas y conquistadores de Córdoba. Ediciones Isquitipe, Buenos Aires

Nores R et al (2017) Diversidad genética en restos humanos arqueológicos del sitio El Diquecito (Costa sur, Laguna Mar Chiquita, Provincia de Córdoba). Rev Argent Antropología Biol 19(1):112

Oliva M (1947) Contribución al estudio de la arqueología del norte de la provincia de Córdoba. Publicaciones Instituto de Arqueología. Linguísticay Folk "Pablo Cabrera" Univ Córdoba (Argentina) 16:3–29

Perez S et al (2009) Discrepancy between cranial and DNA data of early Americans: implications for American peopling. PLoS One 4(5):e5746

Piossek T (1995) Los Hombres de la Entrada. Historia de la expedición de Diego de Rojas 1543–1546. Teresa Piossek Prebisch, Tucumán

Politis G et al (2016) The arrival of *Homo sapiens* into the Southern Cone at 14,000 years ago. PLoS One 11(9):e0162870

Punzi O (1997) Historia de la conquista del Chaco. Epoca colonial. Editorial Vinciguerra, Buenos Aires

Rodriguez J, Ceruti C (1999) In: De Marco D (ed) Nueva historia de la Nación Argentina 1: La Argentina aborigen, conquista y colonización. Editorial Planeta, Buenos Aires, pp 109–122

Serrano A (1945) Los Comechingones. Imprenta de la Universidad Nacional de Córdoba, Córdoba

Chapter 12
A Functional Overview

12.1 Introduction

Although large saline lakes and wetlands share most basic characteristics, it is evident that those occupying different areas around the world under varying climatic and geological situations differ in a set of characteristics that make them unique. Accordingly, each of them requires specific research to achieve an appropriate understanding of their structural and dynamic features, which are needed for the implementation of adequate conservation and sustainable management strategies. In this sense, it was only during the last few decades that research efforts in Mar Chiquita produced enough valuable information on a wide array of disciplines, allowing an initial, comprehensive overview of the main functional aspects of this vast and interesting wetland, which is summarized in the following sections.

12.2 A Saline Wetland Located in a Semi-humid Region

The climatic setting of the Mar Chiquita wetland differs from most saline lakes, usually located in arid regions (Hammer 1986). Placed in a transition between humid and semi-arid climate, the local water balance between annual rainfall and evaporation (on average, 800–900 mm versus 1300–1400 mm) results negative in about 400–500 mm, a value that is clearly smaller than expected for lakes located in more arid areas, where annual rainfall is usually below 400 mm (Hammer 1986). In addition, precipitation in the whole drainage basin is also high as compared with arid regions (Breckle 2002) (Chap. 3). In summary, a wide range of variability in water volume and salinity is a unique characteristic of the Mar Chiquita Lake, which drives key aspects of the lake ecological dynamics.

© Springer Nature Switzerland AG 2019
E. H. Bucher, *The Mar Chiquita Salt Lake (Córdoba, Argentina)*,
https://doi.org/10.1007/978-3-030-15812-5_12

12.3   A Structurally Complex Wetland

The wetland comprises two large subsystems: the Dulce River Wetlands, in the Dulce River final floodplain, and the Mar Chiquita Lake, to the south (Chaps. 1 and 3). These two subsystems are geographically and functionally interconnected through complex hydrological, chemical, and biological processes.

In terms of the system hydrology, the floodplain has a buffer effect on the floods coming from the upper Dulce River before they reach the lake. Firstly, the Dulce River wetland acts as a buffer of the Dulce River overflows before reaching Mar Chiquita Lake. This buffering effect depends on the lake water level, which provides the baseline that controls the flow from the wetland to the lake, given that the area reached by the flood in the wetland not only will be determined by the volume of water released but also will depend on the level of the lake. In other words, the higher the lake level, the larger the flooded area. This factor needs to be considered when modeling flooding events in the Dulce River floodplain (Rodriguez et al. 2006).

Materials transfer between the Dulce River floodplain and the Mar Chiquita Lake through the Dulce River include sediments, organic matter, nutrients, and dissolved pollutants. In addition, a peculiar and important food chain is generated from the river to the lake as a consequence of the large fish biomass that flows from the river into the lake, where fish die rapidly when in contact with the highly saline water. This large biomass of dying fish attracts numerous gulls, herons, and storks to the river delta.

Other less known flow paths of organic matter and nutrients includes organic matter transported by large numbers of flamingos present in the lake that move daily between feeding areas and their breeding colonies in the lake, as has already been shown in wetlands in Spain (Batanero et al. 2017) (Chap. 7). Finally, the massive salt dust storms that occur in Mar Chiquita may also transport substantial amounts of salt between the floodplain and the lake (Bucher and Stein 2016) (Chap. 2).

12.4   Flood and Fire as Key Drivers of the Wetland

In the Dulce River wetland, as in most wetlands, the ecosystem dynamics is shaped by flooding and fire. Flooding provides a sheet flood of water that reaches large areas of terrestrial vegetation and washes salt from soils (Chap. 10). Salt removal by floods is critical since the wetland soils have adverse conditions for plant growth. The wetland soils in these alluvial plains are hydromorphic (developed under flooding conditions), with high salt content and low oxygen availability, which may cause severe imbalances in plant nutrition (Läuchli and Lüttge 2002). In turn, the resulting loss of nutritional quality of pastures may affect mammal herbivores, both wild and domestic. For example, under saline stress, plant growth is reduced, and leaf calcium concentration falls. As a result, cattle grazing in the prairies and Spartina grasslands in Mar Chiquita are frequently affected by calcium deficiency (Bucher and Bucher 2006) (Chap. 10).

Fire, a natural and integral component of the ecosystem, eliminates the dry grass biomass generated by the previous flood and prevents woody vegetation from invading the wetland, maintaining a grass-dominated plant community (Chap. 10). Fire also determines the balance and cycling rate of nutrients, thereby serving as a potent catalyst and limiting factor to the system productivity (Whelan 1995). Unfortunately, the positive effects of flood and fire are somewhat counterintuitive, and therefore not necessarily perceived as beneficial by the local population, which leads to favoring pulse-suppressing policies (Bucher et al. 2006).

12.5   The Influence of the Mar Chiquita Wetland on the Local Climate

Although there is no specific evidence available, it is very likely that the Mar Chiquita wetland (including the lake and the Dulce River wetland) may have a significant influence on the local climate, considering its very large size (of about 10,000 km^2) and the fact that lacking a river outlet, evaporation is the only source of water loss from the wetland (Rodriguez et al. 2006; Troin et al. 2010) (Chap. 3). According to Jørgensen (2010), it could be expected that Mar Chiquita may reduce the region's maximum temperature and increase minimum temperature all year round and also increase precipitation over the lake and nearby areas. There is a clear need for more research on this topic, particularly in relation to future climate change scenarios.

12.6   Salinity as a Critical Driver of the Lake Dynamics

Throughout the twentieth century, water salinity in Mar Chiquita fluctuated between 360 and 25 g/L, which is the most extensive range recorded in a series of salt lakes of the world mentioned by Williams (1993). This extreme variability may be seen as a large-scale natural experiment that provides an opportunity for interesting analyses and comparisons.

Drastic variations in the water volume of the lake and the resulting broad shifts in water salinity lead to substantial changes in the lake energy and nutrient pathways, as well as in the lake biological communities (see Chaps. 3 and 4).

Most of these changes take place at a critical salinity of 55 g/L that corresponds to a lake level of about 68.5 m a.s.l. Obviously, this value represents the middle point of a transitional stage that will vary in amplitude according to different processes and species involved, as mentioned by Reati et al. (1997). The main characteristics of these contrasting situations are described in the following sections.

12.6.1   Lowstands

The lake bottom sediments are characterized by a widespread accumulation of the organic black mud, even at a low depth close to the shore (Chap. 4). Mud accumulation results from sulfur-based oxidative processes under anoxic conditions that are generated by the inverse relationship between salinity and oxygen solubility (Chap. 4). The black mud sediment is an important food source for flamingos (Jenkin 1957) (Chap. 7).

The phytoplankton shows low species diversity, with high dominance of blue-green algae (*Cyanophyceae*) (Chap. 4). The zooplankton is composed almost entirely of single species, the Artemia brine shrimp (*Artemia franciscana*), an algae consumer that may reach a considerable biomass, becoming a principal food source for flamingos (Johnson and Cézilly 2007), shorebirds, and grebes (Jehl 1989) (Chap. 7).

The Ephydra salt fly (Ephydra sp.) is abundant and also an important food source for the Wilson's phalarope (*Phalaropus tricolor*) (Jehl 1989). This fly tends to be commoner at transitional salinities level at the upper end of the lowstand period.

12.6.1.1   Food Chains

The dominant food chains generated in the lake during the lowstands include birds as top consumers. They include the following species:

Chilean flamingo (*Phoenicopterus chilensis*): A species that breed in large colonies in Mar Chiquita. Feeds by filter feeding, mainly on plankton (particularly but not exclusively on Artemia brine shrimp) and also on the abundant black organic mud. The species reaches the highest numbers during the lowstands (Chap. 7).

Wilson phalarope (*Phalaropus tricolor*): Apparently feeds mostly on Artemia brine shrimp and Ephydra salt fly, although more information is needed. During the lowstands, the species reached record numbers in Mar Chiquita (Chap. 7).

Silvery grebe (*Podiceps occipitalis*): Likely to feed mainly on Artemia and Ephydra larvae (no confirming data available).

Gulls, particularly brown-hooded gull (*Chroicocephalus maculipennis*): Very abundant during the lowstands, reaching over a hundred thousand birds. They are scavenger birds that feed mostly on dead animals, mainly fish brought by the tributary rivers that die in contact with the lake's salty water (Chap. 7).

12.6.2   Highstands

During high water periods, when the Mar Chiquita water salinity drops below the threshold concentration of 55 g/L, two very important changes were observed. First, the pejerrey silverside fish (*Odontesthes bonariensis*) invaded the lake. Second, shortly after, until then high population of Artemia brine shrimp (*Artemia franciscana*) dropped to the point of becoming no longer detectable. The disappearance of Artemia was probably due to a salinity level below optimum value for the species and also to high predation pressure by the silverside pejerrey fish. According to Hurlbert et al. (1986), competition between fish and flamingos in the salt lakes of Argentina is mediated by reduction of the invertebrate population (mainly Artemia) by the fish.

Another significant change observed at the onset of the highstand period was the almost complete disappearance of the organic black mud from shallow waters, becoming restricted to deep sites under anoxic conditions. As a result, mud availability to foraging flamingos decreased substantially (Chaps. 4 and 7).

In addition, the phytoplankton diversity increased substantially. Dense littoral submerged plant biomass developed, including filamentous green algae (*Cladophora fracta* and *Mougeotia* sp.), as well as extensive stands of the Beaked tasselweed (*Ruppia maritima*). The zooplankton diversity increased moderately (Bucher and Bucher 2006) (Chap. 4).

12.6.2.1   Food Chains

The dominant food chains during the lowstands weakened markedly during the highstands. The flamingo population dropped (Bucher and Curto 2012). A similar drastic drop was observed in the case of the Wilson's phalarope (Nores 2011). No accurate records are available for the grebe species, but non-systematic observations suggested a drop in the silver grebe population (E.H. Bucher personal observations).

New food chains emerged during this period based on the presence of fish and plants. The pejerrey silverfish showed an exceptional initial population increase, to the point of sustaining a commercial fishing industry (Chap. 5), followed by a slow and steady decline until its extinction in 2005 when water salinity rose again over the 55 g/L threshold. At the same time, the population of piscivorous birds increased substantially, particularly the Neotropic Cormorant (*Phalacrocorax olivaceous*) (Chap. 7).

In addition, availability of plant material led to a marked increase of herbivorous bird populations, including coots (*Fulica* spp.), coscoroba swan (*Coscoroba coscoroba*), and black-necked swan (*Cygnus melancoryphus*), all of them feeding mainly on filamentous green algae and Ruppia tasselweed (*Ruppia maritima*) (see Chaps. 4 and 7).

12.6.3   Recurrence of the High-Low Salinity Cycle

During the 1976–2018 period, the Mar Chiquita Lake went through two cycles of high-low salinity condition, the first from 1978 to 2009 and the second between 2016 and 2018. In both occasions, the sequence of Artemia-silverside dominance and changes in food chains repeated themselves at the same salinity level.

This replication of similar environmental conditions and species response strongly validates the hypotheses that salinity change is a reliable predictor of the described dominant species succession and the associated environmental conditions. In summary, research in Mar Chiquita confirms Hammers (1986) conclusions: "one clear characteristic of saline lakes is that the level of water salinity is the primary factor controlling the biodiversity and functionality of these ecosystems."

References

Batanero GL et al (2017) Flamingos and drought as drivers of nutrients and microbial dynamics in a saline lake. Sci Rep 7:12173

Breckle S-W (2002) Walter's vegetation of the earth. Springer, Berlin

Bucher EH, Bucher AE (2006) Síntesis funcional. In: Bucher EH (ed) Bañados del Río Dulce y Laguna Mar Chiquita. Academia Nacional de Ciencias (Córdoba, Argentina), Córdoba, pp 139–159

Bucher EH, Coria RD, Curto ED, Lima JJ (2006) Conservación y Uso sustentable. In: Bucher EH (ed) Bañados del Río Dulce y Laguna Mar Chiquita. Academia Nacional de Ciencias (Córdoba, Argentina), Córdoba, pp 327–341

Bucher EH, Curto ED (2012) Influence of long-term climatic changes on breeding of the Chilean flamingo in Mar Chiquita, Córdoba, Argentina. Hydrobioloia 697:127–137

Bucher EH, Stein AF (2016) Large salt dust storms follow a 30-year rainfall cycle in the mar Chiquita Lake (Córdoba, Argentina). PLoS One 11(6):e0156672

Hammer UT (1986) Saline lake ecosystems of the world. Dr. W. Junk Publishers, Boston

Hurlbert S et al (1986) Fish-flamingo-plankton interactions in the Peruvian Andes. Limnol Oceanogr 31:457–468

Jehl J Jr (1989) Biology of the Eared Grebe and Wilson's Phalarope in the nonbreeding season: a study of adaptations to saline lakes. Stud Avian Biol 12:1–74

Jenkin P (1957) The filter-feeding and food of flamingoes (*Phoenicopteri*). Philos Trans R Soc Lond B 240:410–493

Johnson AR, Cézilly F (2007) The greater flamingo. T & AD Poyser, London

Jolis J (1972) Ensayo sobre la historia natural del Gran Chaco. Universidad Nacional del Nordeste, Argentina. Instituto de Historia, Resistencia

Jørgensen SE (2010) A review of recent developments in lake modeling. Ecol Model 22(4):689–692

Läuchli A, Lüttge U (2002) Salinity: environment—plants—molecules. Springer, Dordrecht

Nores M (2011) Long-term waterbird fluctuations in mar Chiquita Lake, Central Argentina. Waterbirds 34:381–388

Reati G et al (1997) The Laguna de Mar Chiquita (Córdoba, Argentina): a little known, secularly fluctuating saline lake. Int J Salt Lake Res 5:187–219

Rodriguez A et al (2006) Modelo de Simulación Hidrológica. In: Bucher E (ed) Bañados del Rio Dulce y Laguna Mar Chiquita (Córdoba, Argentina). Academia Nacional de Ciencias (Córdoba, Argentina), Córdoba

Troin M, Vallet-Coulomb C, Sylvestre F, Piovano E (2010) Hydrological modelling of a closed lake (Laguna Mar Chiquita, Argentina) in the context of 20th-century climatic changes. J Hydrol 393:233–244

Whelan RJ (1995) The ecology of fire. Cambridge University Press, Cambridge, UK

Williams WD (1993) The worldwide occurrence and limnological significance of falling water-levels in large, permanent saline lakes. Verh Int Ver Limnol 25(2):980–983

Chapter 13
Conservation and Sustainable Use

13.1 Introduction

This chapter presents a summary of the current conservation status of Mar Chiquita Lake and Dulce River wetland, including three main sections. Firstly, a list of the values that support the conservation of this wetland is presented. Secondly, the environmental threats affecting the region and the key factors that require priority consideration are addressed. Finally, the priorities that should guide conservation actions to strengthen the sustainable use of the Ramsar site.

13.2 The Value of Mar Chiquita

The value assigned to protected areas (such as reserves and parks) has changed over time with the accelerated increase of worldwide environmental degradation and with an increasing number of research results that have revealed new and sometimes complex environmental interactions. At present, valuing criteria for ecosystems includes three basic aspects: scientific and conservation values, environmental services, and economic values they provide to humankind.

13.2.1 Conservation Values

The conservation value of the area is given by its rich biodiversity, which includes several threatened species and some emblematic ones, particularly the three flamingo species found in Mar Chiquita (Chap. 7). Due to its great bird diversity, the vast Mar Chiquita wetland is regarded as an important area for the conservation of the avifauna (IBA) in Argentina (Di Giacomo and Parera 2008).

© Springer Nature Switzerland AG 2019
E. H. Bucher, *The Mar Chiquita Salt Lake (Córdoba, Argentina)*,
https://doi.org/10.1007/978-3-030-15812-5_13

One important ecological value of Mar Chiquita that should receive particular attention is its role as feeding, refuge, and breeding site for many migratory and nomadic bird species. The loss of a salt lake of exceptional value as Mar Chiquita may pose severe threats to the survival of these bird species. This conservation value has been considered one of the primary reasons for the designation of Mar Chiquita as a Ramsar site of international importance. The Ramsar Convention and its partner organizations (Wetlands International, IUCN, BirdLife International, and Worldwide Fund for Nature) and other many small nongovernmental organizations are dedicated to the conservation of salt lakes.

13.2.2   Environmental Services

The most critical environmental services provided by the wetland ecosystems include:

1. Influence on the local climate: the large water mass of the Mar Chiquita Lake buffers temperature extremes around the lake and increases local rainfall (Chap. 3).
2. Contribution of saline wetlands like Mar Chiquita to the completion of the nitrogen cycle by releasing elemental nitrogen into the atmosphere, helping to decrease pollution from excessive use of fertilizers in nearby agricultural areas.
3. Sequestration of atmospheric carbon by the phytoplankton that remains immobilized in the lake sediments (Chap. 4).
4. Immobilization of toxic substances (including agrochemicals and heavy metals) in the lake sediments (Chap. 4).
5. Diminished release of methane to the atmosphere. Sulfate-rich sites such as salt wetlands like Mar Chiquita release smaller methane fluxes than low-sulfate wetlands in the same locality (Gauci et al. 2004) (Chap. 4).
6. Amelioration of exceptional floods produced by high flows from the Dulce River before reaching the Mar Chiquita Lake (Chap. 3).

13.2.3   Scientific Value

The relative ecological simplicity and discreteness of salt lakes are not the only reasons why these water bodies are interesting to ecologists. In fact, these systems are complex since it is possible to regard salt lakes as a continuous series of ecosystems, from simple (the most saline) to complex (the least saline) for the study of ecosystem attributes. In other words, they can be viewed as large-scale experimental situations that undergo permanent variations (Jellison et al. 2008). Moreover, the sedimentary series present in the lake bottom provide valuable information regarding the geological history of the Mar Chiquita region. A review of the scientific research developed in Mar Chiquita is given in Bucher (2006) and in the present book.

13.2.4   Recreational and Aesthetic Values

Due to Mar Chiquita's attractive landscape and diverse and abundant birds and other wildlife, the area has become a popular tourist destination, with a continually increasing number of visitors. In addition, therapeutic tourism based on treatments with the organic black mud (Chap. 4) was popular during the lowstand period. Even if the curative properties of the black mud remain indeterminate, considerable credence was given to its therapeutic values around the world (Williams 1998). Availability of the black mud could return in case the lake water level lowers back to a high-salinity situation.

13.2.5   Economic Values

The present predominant economic activity in the wetland is tourism based on the recreational and aesthetic values of the wetland, as described previously. This activity is growing steadily, particularly in the only coastal resort, Miramar. As mentioned previously, therapeutic tourism was an additional attraction during the lowstand period.

During the highstand period that started in the 1970s, the invasive pejerrey silverside fish has been a potential, although intermittent, resource, which depends on the lake water level and salinity. By contrast, the Artemia brine shrimp (*Artemia franciscana*) is a potential resource (never exploited in this lake) during the lowstand periods with high water salinity (Chap. 4). In both cases, the potential extractive exploitation of these species needs to be confronted within the limits imposed by the protected area and Ramsar site status of the Mar Chiquita Lake.

Sport hunting tourism in the wetland area (primarily doves and ducks and other wildlife to a lesser extent) is not allowed in the Province of Córdoba but still legal in the province of Santiago del Estero. This activity attracts a large number of international tourists. Use of lead ammunition is still allowed, despite its negative environmental impact. Sport fishing tourism is growing and has considerable potential along the lower and middle courses of the Dulce River.

With regard to extractive uses, in the past wildlife trapping for the fur industry was very important in Mar Chiquita, including mainly the coypu, locally known as nutria (*Myocastor coypus*), and other wild mammals (mainly foxes, cats, and reptiles (particularly the iguana lizards (*Tupinambis* spp.) and the lampalagua boa snake (*Boa constrictor*))) (see Chaps. 6 and 8). This activity has ceased entirely due to legal restrictions and a lack of demand.

Land use in the Dulce River wetlands was traditionally based on cattle, goat, and sheep production. Cattle management was based on transhumant pastoralism, which consists of moving the animals selecting the grazing areas according to the flooding of the Dulce River. At present this practice is being rapidly replaced by sedentary agriculture and cattle raising, which brings critical environmental conflicts (see Sect. 13.2).

Mineral salts, particularly evaporites such as sodium chloride and sodium sulfate, are potential resources. However, they have never been exploited at a commercial scale, mainly because concentration in the salt deposits in Mar Chiquita is lower (and therefore less profitable) than other deposits in Argentina. Lithium is not present in a concentration level that may justify commercial exploitation (Durigneaux 1978). It is also possible that other valuable natural products may be discovered in the future, which may require further research efforts. For example, research on bacteria capable of adapting to extreme environments (extremophiles) appears promising (Oren 2002).

13.3   Threats

The main threats affecting the Mar Chiquita region include (a) water diversion and alteration of the basin hydrological regime, (b) habitat alteration and changes in land use, and (c) pollution. In turn, each one of those threats may produce a cascade of additional negative impacts.

13.3.1   Water Diversion and Alteration of the Hydrological Regime

The Mar Chiquita drainage basin, as in the entire South American Gran Chaco ecoregion, is affected by a marked process of agricultural frontier expansion that implies additional water demand, while, at the same time, drainage basins are affected by deforestation and massive erosion. Water diversion is a widespread threat to wetlands and particularly to saline lakes. Unfortunately, the worldwide extent of salt lake desiccation has received scant attention, despite the fact that some cases have become internationally well known, such as the Dead Sea, the Aral Sea in Asia, and the Mono Lake in the United States (Jellison et al. 2008; Shiklomanov 2000; Wurtsbaugh et al. 2017).

In contrast to this global trend to desiccation of salt lakes, the Mar Chiquita vast wetland is subjected to a peculiar situation of contradictory characteristics. Since the 1970s, the water level has extraordinarily increased, with unprecedented previous records in its geological history, due to a marked increase in the regional precipitation regime (see Chaps. 3 and 4).

Despite this highstand situation, which is clearly associated with global climate change, the most serious environmental threat to Mar Chiquita continues to be water diversion from the tributaries, particularly the Dulce River. The expansion of irrigated agriculture and water consumption for urban use in the middle and upper courses of the Dulce River show a sustained increasing trend, as detailed below.

Water withdrawal for urban consumption involves three important cities, with population sizes of 1,600,000 (Tucuman), 900,000 (Santiago del Estero), and 45,000 (Termas de Rio Hondo) inhabitants, plus several smaller urban and rural

Fig. 13.1 Massive fish mortality caused by mismanagement of the irrigation channel interconnection that resulted in living a lagoon totally dry in a few hours. (Image from Bucher et al. 2006)

areas. The area under irrigation for agriculture was 66,000 ha in Tucuman and 107,000 ha in Santiago del Estero in 2010 (Procisur 2010). In total, the present water diversion between the Rio Hondo dam and Mar Chiquita reaches about 1080 hm^3/year and is expected to add 190 hm^3/year in the next few years (Gallego 2012).

In addition to water withdrawal, increasing flow regulation interferes with the annual floods reaching the Dulce River wetland and may have severe environmental effects on vegetation productivity and ecosystem integrity of the Dulce River wetlands (see Chaps. 3 and 10). At present, there are three large dams along the Dulce River (Escaba, and El Cadillal in Tucumán, and Rio Hondo in Santiago del Estero), and more are being planned (Gallego 2012). Flow regulation implies other risks. For example, mismanagement of the irrigation system in lower Dulce River has caused several incidents of fish mortality due to the sudden deviation of water to irrigation channels, leaving large masses of fish trapped with no water in small branches and lagoons (Fig. 13.1) (Bucher et al. 2006).

13.3.2 Coast Conservation and Management

Mar Chiquita has an extended coastline of about 300 km, which fluctuates according to the lake level (Chap. 4). This coastline is at present almost empty of human urban centers, with the exception of the only coastal city (Miramar) (Chap. 1).

However, this situation may change rapidly in the near future, taking into consideration the rapid growth and development of the region that surrounds the Mar Chiquita wetland.

At present, the primary conservation problem of the Mar Chiquita coastal area derives from the rapid deforestation of the native Chaco woodland that originally covered most of the lake perimeter. This forest has been replaced to a large extent by agricultural land (Fig. 13.2). Conservation of the few remaining fragments and restoration of additional areas wherever possible deserve high priority.

Another equally important conservation priority is wildlife protection, since many species found feeding, breeding, and resting habitat in the shore area, as, for example, the many species of migratory shorebirds that visit Mar Chiquita (Chap. 7). Additional conservation challenges include the need for establishing adequate protection measures for the valuable archaeological, paleontological, and geological sites dispersed along the Mar Chiquita coastal zone.

13.3.3   Changes in Land Use in the Rio Dulce Wetland

A new drastic and sudden case of land-use change is taking place in the Dulce River wetland since early in the twenty-first century when new land-use practices were introduced in the region. These changes followed the arrival of new investors, in some cases with property titles of dubious validity, in a process frequently referred to as "land grabbing" (Suliema 2018). The traditional transhumant livestock management was replaced with an intensive, sedentary management system that resulted in the loss of native vegetation and the extensive fencing of the area and also had significant social and environmental adverse effects (Chap. 11) (Bucher 2016).

13.3.4   Pollution

Although pollution problems are in general limited in the sparsely populated Mar Chiquita wetland, there are indications that this situation is changing. Pollution problems in the Mar Chiquita region may be grouped into two main categories: those originated from pollutants transported by the tributary rivers and those generated locally.

The tributary rivers have the common feature that all them flow across areas under intensive agriculture, including irrigated lands, where they receive great loads of agrochemicals. At present, the Primero River is clearly the primary source of contaminants entering the Mar Chiquita Lake. This river reaches Mar Chiquita with a high load of pollutants, mostly from urban waste from Córdoba city (with a population of 1,400,000 inhabitants), which are not appropriately treated (Wunderlin 2018). Inorganic substances, including heavy metals, as well as pharmaceuticals and organic compounds have been detected along the Primero River course, reaching Mar Chiquita (Gaiero et al. 1997; Stupar et al. 2014).

Fig. 13.2 (**a**) Fragment of the original Chaco woodland that covered most of the Mar Chiquita coasts. (**b**) The original forest has been replaced to a great extent by an agriculture-dominated landscape

Among inorganic contaminants, arsenic poses a significant problem to public health in the whole Chaco Pampas region that includes Mar Chiquita, since this is one of the largest regions in the world with high levels of arsenic in groundwater. In general, arsenic concentrations in surface waters are markedly lower than those in groundwater. Stupar (2013) found that arsenic concentrations in the Primero River were below the World Health Organization drinking water guideline (22.4 µg/L) at most of the sampling points, except in those located close to Mar Chiquita, where arsenic concentration was about twice those the drinking water guideline. Arsenic concentration in groundwater is a primary cause of concern regarding drinking water supply for most urban centers around Mar Chiquita (Nicolli et al. 2012). Regarding mercury, Stupar et al. (2014) found mercury concentrations along the Primero River down to its mouth in the Mar Chiquita Lake ranging between ~13 and ~131 µg/kg, indicating a system with low pollution.

Organic contaminants were assessed by Ballesteros and Bistoni (2014), who sampled water, sediments, and silverside fish (*Odontesthes bonariensis*) tissues in the mouth of the Primero River in Mar Chiquita. Their results showed the occurrence of organochlorine pesticides (OCPs), polychlorinated biphenyls (PCBs), and polybrominated diphenyl ethers (PBDEs). Endosulfans, DDTs, PCBs, and PBDEs were found in all sampling sites. OCPs/(PCBs + PBDEs) > 1 ratios indicated the dominance of agriculture over industrial and urban pollution sources. The level of γ-HCH and endosulfan OCPs exceeded national and international quality guidelines of water and sediments, which represent a risk to the aquatic biota. Ballesteros and Bistoni (2014) also found that contaminants in the pejerrey silverside tissues showed dominance of endosulfan (used in agricultural areas around the lake) and PCBs, industrial pollutants probably transported by the Primero River from the Córdoba City area. PCB levels in the silverside fish muscle exceeded the ADI (allowing human consumption) limit, being therefore a risk for human health.

The Dulce River receives a high contamination load from local industrial and agricultural effluents that discharge into the Rio Hondo dam, which is highly polluted (Bucher et al. 2007). From there, the river flows across areas with high development of irrigated agriculture, which contribute agrochemicals to the river. Further south, the river flows across a naturally vegetated and barely populated area along about 100 km, which may reduce its pollutant load at the point of reaching Mar Chiquita. Unfortunately, no systematic monitoring is being performed.

The main sources of locally generated pollution include urban wastes, agrochemicals products, and lead from hunting and fishing activities. Release of urban pollutants in the lake is still limited to the only coastal locality, Miramar, still small (3500 inhabitants), but growing fast. Pollution from agriculture sources is very likely, considering that the wetland is surrounded by land under intensive agriculture production and supported by results reported by Ballesteros and Bistoni (2014). Unfortunately, this problem is not systematically monitored. Nevertheless, episodes of massive mortality of birds caused by pesticides have already been recorded in Mar Chiquita (Bucher et al. 2006). The best-known case that attracted worldwide attention was the massive insecticide-induced mortality of the Swainson's hawk *Buteo swainsoni* in the Pampas region of Argentina (including Mar Chiquita) in 1995–1996 (Goldstein et al. 1999).

Lead pollution from hunting and fishing is another important cause of concern in the Mar Chiquita wetlands. The two main lead sources include lead shot used for hunting waterfowl and fishing sinkers and jigs (Esler 1988). Lead shot is no longer used in the portion of the Mar Chiquita wetland under the Córdoba jurisdiction but is still allowed in Santiago del Estero. Lead shots may kill birds directly by ingestion or indirectly when birds of prey or scavenger animals feed on dead birds with lead shot particles in the flesh (Newton 2016). Evidence of lead shot pellets in the flesh of live birds in the Mar Chiquita region has already bee reported by Ferreyra et al. (2014).

13.4 Management Problems and Needs

Management of a vast wetland as Mar Chiquita is very demanding in terms of planning, implementation, and monitoring. Key management challenges include the following:

1. *Ensuring sufficient water resources.* As mentioned before, the Mar Chiquita wetland requires adequate water flow both in terms of the amount of water required and the need for maintaining the annual flood pulse that ensures the ecological integrity of the system. At present, the use of the water resource in the Dulce River drainage basin is regulated by the Inter-provincial Agreement signed by the provinces that share the Dulce River basin in 1967 (Bucher et al. 2006).

   This agreement has severe limitations, the most important one being that it does not ensure a minimum flow needed to preserve the wetland integrity, even during dry years (see Rodriguez et al. (2006) for more details). Clearly, a new management plan for the whole Dulce River catchment basin is urgently needed, taking into account that further increases in water deviation are to be expected, as already mentioned (Chap. 3).

2. *Integration and cooperation between provincial governments.* The fact that the Mar Chiquita Wetland is shared by two provinces poses difficulties in terms of coordination and effectiveness in the management of the protected area, including legal conflicts, species-related management, conservation problems, tourism, etc.

3. *Addressing the lack of adequate human resources and infrastructure.* Given the extension and inaccessibility of some areas in the wetland, well-trained human and infrastructure resources are essential, but usually not available in the required magnitude.

4. *Develop and implement a coastal management plan for Mar Chiquita.* The nature of the challenges to be faced demands that "passive" conservation measures such as creation of reserves be supplemented by active management practices. An ecosystem-wide approach to management would be required to achieve conservation goals, especially maintenance of habitat diversity along the lake's coasts.

References

Ballesteros M, Bistoni M (2014) Multimatrix measurement of persistent organic pollutants in Mar Chiquita, a continental saline shallow lake. Sci Total Environ 490:73–80

Bucher EH (ed) (2006) Bañados del Rio Dulce y Laguna Mar Chiquita (Córdoba, Argentina). Academia Nacional de Ciencias (Córdoba, Argentina), Córdoba

Bucher E (2016) El futuro incierto de los humedales del Chaco: el caso de los Bañados del Río Dulce. Paraquaria Nat 4(2):11–18

Bucher E, Coria R, Curto E, Lima J (2006) Conservación y uso sustentable. In: Bucher E (ed) Bañados del Rio Dulce y Laguna Mar Chiquita (Córdoba, Argentina). Academia Nacional de Ciencias (Córdoba, Argentina), Córdoba, pp 327–341

Bucher EH, Echevarria AL, Chani JM (2007) In: Cicerone D, Hidalgo M (eds) Los humedales de la cuenca del Rio Sall, Argentinal. Jorge Baudino Ediciones, Buenos Aires

Di Giacomo A, Parera A (2008) Veinte áreas prioritarias para la conservación de las aves migratorias en los pastizales del cono sur de Sudamérica. Aves Argentinas, Buenos Aires

Durigneaux J (1978) Composición química de las aguas y barros de la Laguna Mar Chiquita en la provincia de Córdoba. Academia Nacional de Ciencias (Córdoba, Argentina), Córdoba

Esler R (1988) Lead hazards to bird, fish, wildlife, and invertebrates: a synoptic review. U.S. Fish and Wildlife Service, Washington, DC

Ferreyra H, Romano M, Beldomenico P, Caselli A, Correa A, Uhart M (2014) Lead gunshot pellet ingestion and tissue lead levels in wild ducks from Argentine hunting hotspots. Ecotoxicol Environ Saf 103:74–81

Gaiero D, Roman Ross G, Depetris PJ, Kempe S (1997) Spatial and temporal variability of total non-residual heavy metals content in stream sediments from the Suquia River system, Córdoba, Argentina. Water Air Soil Pollut 93:303

Gallego A (2012) Santiago del Estero y el agua: crónica de una relación controvertida. Lucrecia Editorial, Santiago del Estero

Gauci V, Matthews E, Dise N, Walter B, Koch D, Granberg G, Vile M (2004) Sulfur pollution suppression of the wetland methane source in the 20th and 21st centuries. Proc Natl Acad Sci 101(34):12583–12587

Goldstein M, Lacher T, Woolbridge B, Bechard M, Canavelli S, Zaccagnini M, Gobb JR, Hooper MJ (1999) Monocrotophos-induced mass-mortality of Swainson's hawk in Argentina1995–1996. Ecotoxicology 8:201–214

Jellison R, Williams WD, Timms B, Alcocer JD, Aladin N (2008) Salt lakes: values, threats and future. In: Polunin NVC (ed) Aquatic ecosystems: trends and global prospects. Cambridge University Press, Cambridge, UK, pp 94–110

Newton I (2016) Problems created by the continuing use of lead ammunition in game hunting. Br Birds 109:172–179

Nicolli HB, Bundschuh J, Blanco MC, Tujchneider OC, Panarello HO, Dapeña C, Rusansky JE (2012) Arsenic and associated trace-elements in groundwater from the Chaco-Pampean plain, Argentina: results from 100 years of research. Sci Total Environ 429:36–56

Oren A (2002) Halophilic microorganisms and their environments. Kluver Academic Publishers, London

Procisur (2010) El riego en los países del Cono Sur. IICA, Montevideo

Rodriguez A, Pagot M, Hillman G, Pozzi C, Plencovich G, Caamaño Nelli G, Oroná C, Curto E, Bucher E (2006) Modelo de simulación hidrológica. In: Bucher E (ed) Bañados del Rio Dulce y Laguna Mar Chiquita (Córdoba, Argentina). Academia Nacional de Ciencias (Córdoba, Argentina), Córdoba

Shiklomanov IA (2000) Appraisal and assessment of world water resources. Water Int 25:11–32

Stupar YV (2013) Trends and rates of mercury and arsenic in sediments accumulated in the last 80 years in the climatic-sensitive Mar Chiquita system, Central Argentina. Université Sciences et Technologies, Bordeaux

Stupar Y, Schäfer J, García MG, Schmidt S, Piovano E, Blanc G, Huneaue FP, Le Coustumera P (2014) Historical mercury trends recorded in sediments from the Laguna del Plata, Córdoba, Argentina. Chem Erde Geochem 74(3):356–363

Suliema HM (2018) Exploring the spatio-temporal processes of communal rangeland grabbing in Sudan. Pastoralism 8:14

Williams WD (1998) Management of inland saline waters. United Nations Environmental Programme (UNEP)

Wunderlin DA (2018) The Suquía River Basin (Córdoba, Argentina). An integrated study on its hydrology, pollution, effects on native biota and models to evaluate changes in water quality. Springer, Cham

Wurtsbaugh WA, Miller C, Null SE, DeRose R, Justin Wilcock P, Hahnenberger M, Howe F, Moore J (2017) Decline of the world's saline lakes. Nat Geosci 10(11):816–821

Index

Printed in the United States
By Bookmasters